Fundamentals of

FOURIER TRANSFORM INFRARED SPECTROSCOPY

Fundamentals of FOURIER TRANSFORM INFRARED SPECTROSCOPY

Brian C. Smith

CRC Press
Boca Raton New York London Tokyo

Library of Congress Cataloging-in-Publication Data

Smith, Brian C.
Fundamentals of Fourier transform infrared spectroscopy / by Brian C. Smith.
p. cm.
Includes bibliographical references (p. –) and index.
ISBN 0-8493-2461-0
1. Fourier transform infrared spectorscopy. I. Title.
QD96.I5S59 1996
543'.08583--dc20 95-44659
CIP

International Standard Book Number 0-8493-2461-0
Library of Congress Card Number 95-44659
Printed in the United States of America 2 3 4 5 6 7 8 9 0
Printed on acid-free paper

Preface

The history of Fourier Transform Infrared (FTIR) spectroscopy over the last 25 years has been remarkable. Starting with a few commercial instruments in the late 1960s, the technique has grown so popular as to almost totally supplant dispersive instruments. that were the bedrock of infrared analysis since the 1940s. Through the use of FTIR, sensitivity has increased, the size of the smallest sample that can be analyzed has been reduced, and the number of new applications has exploded. My reasons for writing this book are twofold. First, to give a basic introduction to FTIR. Second, to update information on older techniques and discuss newer techniques that did not even exist 10 years ago.

The intended audience for this book is people who are new to FTIR, but who may already have some technical background. As the utility and importance of FTIR has increased, more non-spectroscopists are becoming interested in applying the technique to their own fields. Also, due to shake-ups in the employment market, people are finding themselves "inheriting" instruments they do not know how to use, or are being transferred into jobs whey they are expected to perform FTIR analyses. All of these people are in need of detailed technical information about FTIR so they can perform their jobs. Additionally, this book could be used as a supplemental text in undergraduate and graduate level instrumental analysis courses.

I have endeavored to introduce FTIR assuming little or no background knowledge on the part of the reader, and have kept the mathematics to a minimum. A nodding familiarity with freshman chemistry and physics would be useful in reading this book, but is by no means required. By using words and pictures I have tried to make difficult mathematical concepts such as the Fourier transform easier to understand. In addition to discussions of theory and instrumentation, the book is full of practical hands-on tips on how to use an FTIR quickly and effectively. These "tricks of the trade" were culled from my 15 years of experience in the field, and from discussions with many FTIR experts. These are the kinds of details new FTIR users need to quickly get up to speed on performing analyses correctly.

The book is organized as follows. After a brief introduction in Chapter 1, Chapter 2 tackles the important concepts behind how an FTIR works. The idea here is to teach students about the inside of the instrument straightaway, and get them up to speed on the jargon of the field so that discussions in later chapters can assume familiarity with the terms. Chapter 3 discusses the theory behind how spectra are manipulated using computers. The easy availability of spectral manipulation programs, and the lack of guidance as to how to use these programs, can cause users to unwittingly add artifacts or destroy peaks in good spectra. This chapter's emphasis on following specific guidelines for spectral manipulations is, I believe, a unique approach to the problem.

Chapter 4 is on sampling techniques. Since entire books have been written on infrared sampling, my treatment of the topic is not as detailed as possible. The emphasis is on what types of sampling work best for which samples. My experience has been that using the right sampling technique is half the battle in obtaining good spectra. The focus has been on practical hands-on advice that everyday practitioners of FTIR will find useful. Theory is discussed where appropriate.

Chapter 5 is on quantitative analysis, one of the most important uses of FTIR. Again, an entire book could have been written on this topic, but the emphasis has been on discussing the basics of simple single component analysis. The chapter finishes with a semimathematical discussion of the different techniques used in multicomponent quantitative analysis. The final chapter of the book discusses more advanced techniques, such as infrared microscopy, and interfacing an FTIR to gas and liquid chromatographs. This information was included because these techniques are either new, important, or have changed radically in the last few years. Chapter 6 is followed by a glossary of important FTIR terms. Throughout the text, important terms are denoted *by italics*, and the defintions of most of these terms appear in the glossary.

In conclusion, I hope you find this book a readable and enjoyable introduction to an important field of chemical analysis. I take full responsibility for the content of the book, including all errors and omissions. If you have any questions or comments on the material in this book, please contact me.

Happy Reading,

Brian C. Smith 10/19/95

Brian C. Smith, Ph.D
Spectros Associates
334 Mendon Rd.
Northbridge, MA 01534
E-mail: Spectros1@aol.com

Acknowledgements

I have been privileged throughout my career to have had the advice and encouragement of many people. I have also had the opportunity to interact with many fine people during the creation of this book. I would like to take this opportunity to thank the many people who influenced my career, and who contributed to the completion of this book.

There are several people responsible for my being an infrared spectroscopist in the first place. My high school chemistry teacher, Gerald Nesbitt, made learning chemistry great fun and convinced me to pursue chemistry as a career. Dr. George Takacs of Rochester Institute of Technology made physical chemistry exciting and understandable to me, and influenced me to pursue graduate work in that field. My baptisimal into the infrared world occurred while working with Dr. William Golden of IBM, and my introduction to FTIR was while working at Bell Labs with Dr. Elsa Reichmanis. My graduate work in FTIR was performed with Dr. John Winn of Dartmouth College, whose patience and technical knowledge enabled me to learn a great deal.

A large number of the figures and spectra in this book were prepared using the GRAMS/386 software package created by Galactic Industries of Salem, NH. I would like to thank the people there for their advice on how to use the software. They are Don Keuhl, Jamie Duckworth, Tony Nip, Dave Manning, and Barbara Marschak.

A number of people have materially contributed to this book, either by making specific suggestions, or by giving me artwork and spectra, or by proofreading chapters or the entire volume. I owe them a debt of gratitude. They are Dr. Joseph "Opie" Van Gomple of Analytical Solutions, Dr. Laurie Sparks of Westinghouse Hanford Technical Services, Phyllis Bender, Joan Panagos, Jeff D'Agostino, and Cindy Baulsir of Spectra-Tech, Jim Dwyer of Lab Connections, and Dr. J. McClelland of MTEC Photoacoustics.

This book has been used as a training manual for the course "Fundamentals of FTIR" taught by me for Spectros Associates, and has also been used as part of the American Chemical Society's FTIR short course that I teach. Over 1,000 people have read and commented on the book, and it has been improved greatly thanks to their comments and criticisms. I am very grateful to my course attendees for their honesty and help in this endeavor.

I thank my "partner in crime" here at Spectros Associates, Peggy Veal. Her organization, commitment, and knack for handling a million little details have helped immensely in freeing me up to work on this book. I also thank her for being a sounding board for my crazy business ideas, and for being a good friend.

Most of all, I would like to thank Marian Ellwood for her love, warmth, and support in this endeavor, and in our life together.

In conclusion, and for safety's sake, I would like to thank the entire Eastern and Western Hemispheres.

Table of Contents

Dedication

This book is dedicated to the memory of my grandparents

Louis Whitney Smith
Lula Pearl Hardy Smith

Donald Lee Blue
Mildred Julia Ewart Blue

and to the memory of my dear aunt

Elizabeth Blue Fisher

Chapter 1

Introduction to Fourier Transform Infrared Spectroscopy

A. Terms and Definitions

The purpose of this book is to introduce the reader to the fundamental concepts of Fourier Transform Infrared (FTIR) spectroscopy. The discussion assumes no previous background in FTIR, but a familiarity with the basic concepts of chemistry and physics will be helpful in understanding the text. The book teaches the basics of FTIR to those new to the field, and will serve as an excellent reference guide for experienced users. All terms shown in *italics* will be defined in the glossary at the end of the book.

Infrared Spectroscopy is the study of the interaction of infrared light with matter. Light is composed of electric and magnetic waves. These two waves are in planes perpendicular to each other, and the light wave moves through space in a plane perpendicular to the planes containing the electric and magnetic waves. It is the electric part of light, called the *electric vector,* that interacts with molecules. The amplitude of the electric vector changes over time and has the form of a sine wave, as shown in Figure 1.1. The *wavelength* of a light wave is the distance between adjacent crests or troughs. Throughout the rest of this book, wavelength will be denoted by the small Greek letter lambda (λ). The *wavenumber* of a light wave is defined as the reciprocal of the wavelength, or

$$W = 1/\lambda$$

where W is wavenumber. If λ is measured in cm, then W is reported as cm^{-1}, sometimes called reciprocal centimeters. W is really a measure of the number of waves (counted as the number of crests or troughs) there are in a centimeter. Wavenumbers are the units typically used in infrared spectroscopy to

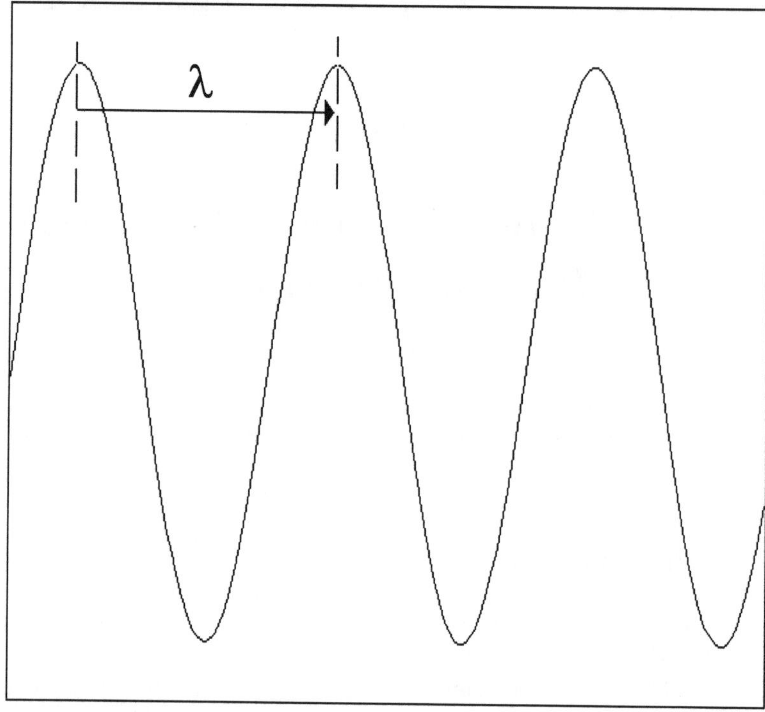

Figure 1.1 An example of a light wave. The arrow denotes the distance between two consecutive crests in the wave, known as a *wavelength.*

denote different kinds of light. The wavenumber of a light wave is directly proportional to energy as follows:

$$E = hcW$$

where:

E = Light energy
c = The speed of light (3×10^8 meters/second)
h = Planck's constant (6.63×10^{-34} Joule-second)
W = Wavenumber

Thus, high wavenumber light has more energy than low wavenumber light. For the purposes of this book, *mid-infrared radiation* will be defined as light between 4000 and 400 cm^{-1}. The majority of FTIRs operate in this wavenumber range.

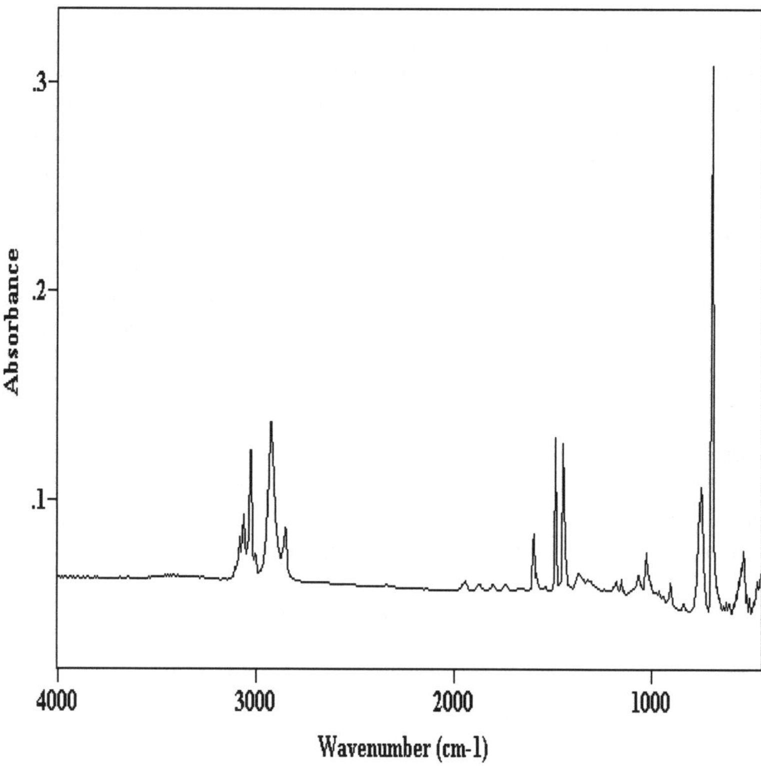

Figure 1.2 An example of an infrared spectrum. This is the infrared spectrum of polystyrene.

Infrared radiation is another name for heat. All objects in the universe at a temperature above absolute zero give off infrared radiation. When infrared radiation interacts with matter it can be absorbed, causing the chemical bonds in the material to vibrate. The presence of chemical bonds in a material is a necessary condition for infrared absorbance to occur. Chemical structural fragments within molecules, known as *functional groups*, tend to absorb infrared radiation in the same wavenumber range regardless of the structure of the rest of the molecule that the functional group is in. For instance, the C=O stretch of a carbonyl group ocurrs at ~ 1700 cm^{-1} in ketones, aldehydes, and carboxylic acids. This means there is a correlation between the wavenumbers at which a molecule absorbs infrared radiation and its structure. This correlation allows the structure of unknown molecules to be identified from the molecule's infrared spectrum. This is what makes infrared spectroscopy a useful chemical analysis tool.

A plot of measured infrared radiation intensity versus wavenumber is known as an *infrared spectrum*. As an example, the infrared spectrum of polystyrene is shown in Figure 1.2. Traditionally, infrared spectra are plotted with high wavenumber on the left and low wavenumber on the right. This means that as a spectrum is read from left to right one is looking from high energy to low energy. Note in Figure 1.2 that the Y axis is in absorbance, and that the X axis scale is in wavenumber. The upward pointing peaks represent wavenumbers at which the sample absorbed infrared radiation. The Y axis can also be plotted in transmittance, in which case the peaks would point down and represent wavenumbers where the sample transmitted less infrared radiation. Regardless of the Y axis units, it is the wavenumber (X axis) positions of these peaks that correlates with molecular structure. For instance, it is well known that the peaks around 3000 cm^{-1} in the infrared spectrum of polystyrene are due to CH bond stretching.

In addition to chemical structures, infrared spectra can provide quantitative information as well, such as the concentration of a molecule in a sample. The basis of all quantitative analyses in FTIR is Beer's law, which relates concentration to absorbance, and has the following form:

$$A = \varepsilon\, l\, c \quad (1.1)$$

where:

A = Absorbance
ε = Absorptivity
l = Pathlength
c = Concentration

The absorbance is measured as a peak height, peak height ratio, peak area, or peak area ratio from the FTIR spectrum. The absorptivity is the proportionality constant between concentration and absorbance. It changes from molecule to molecule, and from wavenumber to wavenumber for a given molecule. For example, the absorptivity of acetone at 1700 cm^{-1} is different than the absorptivity for acetone at 1690 cm^{-1}. However, for a given molecule and wavenumber, the absorptivity is a fundamental physical property of the molecule. For example, the absorptivity of acetone at 1700 cm^{-1} is as invariant as its boiling point or molecular weight. The units of ε are usually given in (concentration x pathlength)$^{-1}$, so the absorptivity cancels the units of the other two variables in Beer's law, making absorbance a unitless quantity. The width of an infrared band gives information about the strength and nature of molecular interactions. Thus, an infrared spectrum provides a great deal of information about a sample.

An instrument used to obtain an infrared spectrum is called an *infrared spectrometer*. There are several kinds of instrument used to obtain infrared spectra. The rest of this book will be devoted to a discussion of Fourier Transform Infrared (FTIR) instruments. After this introductory chapter the topics to be covered include how an FTIR works, how to properly manipulate infrared spectra, single-component and multicomponent quantitative analysis, how to choose the appropriate sampling technique, and interfacing FTIRs with other analytical chemical equipment to obtain more information about a sample.

B. A Brief History of FTIR

A complete understanding of any subject is not possible without an appreciation of the historical events that contributed to its development. Only by knowing where FTIR has come from can we truly understand where it is today, and predict where it will head in the future. The development of FTIR would have been impossible without the use of the Michelson interferometer. This optical device was invented in 1880 by Albert Abraham Michelson [1, 2]. He was awarded the Nobel Prize in Physics in 1907 for accurately measuring wavelengths of light using his interferometer. Michelson was the first American to win a Nobel prize, and was instrumental in establishing the United States as a first-rate scientific power.

The Michelson interferometer was not originally designed to perform infrared spectroscopy. It was designed to test the existence of a "luminiferous aether", a medium through which light waves were thought to propagate. In the famous Michelson-Morley experiment, Michelson used his interferometer to show there is no evidence for the existence of a luminiferous aether. This prompted the questioning of the entire foundation of physics up to that time, and eventually lead to Einstein's discovery of special relativity. In a sense, the Michelson-Morley experiment produced the most famous "negative" result in the entire history of science.

Dr. Michelson was aware of the potential use of his interferometer to obtain spectra, and manually measured many interferograms. Unfortunately, the time consuming calculations required to convert an interferogram into a spectrum made using an interferometer to obtain spectra impractical. The invention of computers made calculating Fourier transforms faster. However, this was not enough. Advances in how computers perform mathematical operations were also necessary to make FTIR a reality. The major advance in this area was made by J.W. Cooley and J.W. Tukey, then at Bell Labs, who invented what is known as the "Fast Fourier Transform" (FFT), or "Cooley-Tukey Algorithm" [3]. This algorithm quickly performs Fourier transforms on a computer, and is still the basis for the transformation routines used in commercial FTIRs. The marriage of the FFT algorithm and minicomputers was the breakthrough that made FTIR possible.

The first commercially available FTIRs were manufactured by the Digilab subsidiary of Block Engineering in Cambridge, Massachusetts during the late 1960s [4, 5]. The efforts of Peter Griffiths, Raul Curbelo, L. Mertz and others made this possible. These instruments incorporated many features we now take for granted in an FTIR: the use of an air bearing driven Michelson interferometer, a He-Ne reference laser used to track mirror position and act as a wavelength standard, and a minicomputer running the Cooley-Tukey algorithm. These instruments made possible the acquisition of quality, high resolution data in a short period of time, and established the advantages of FTIR over previous means of obtaining infrared spectra.

Since the 1960s many other companies have begun manufacturing and selling FTIRs in the United States. The mid-1970s saw the entry of Nicolet Instruments of Madison, Wisconsin into the FTIR business, who quickly became one of the largest FTIR manufacturers. The early 1980s saw the entry of instrument company giant Perkin-Elmer Corp. of Norwalk, Connecticut into the FTIR business. Perkin-Elmer helped found the field of commercial infrared spectroscopy in the late 1940s with what are known as dispersive spectrometers (see below). Their FTIRs were amongst the first to be used in quality control labs due to their low cost, and opened up a whole new market for FTIR.

In 1983, a Nicolet employee named Dave Mattson left that company to start his own FTIR business, Mattson Instruments. His company pioneered the use of the cube corner interferometer and personal computers for FTIR data acquisition and processing. Several other companies, including KVB/Analect (Irvine, CA), Midac (Irvine, CA), Bomem (Chicago, IL), and Bruker (Billerica, MA) have also contributed to the commercial development of FTIR.

The advantages of FTIR quickly gave rise to the adaptation of attenuated total reflectance (ATR) [6] and diffuse reflectance (DRIFTS) [7] for the rapid and reproducible analysis of liquids and solids. The combination of gas chromatography (GC) and FTIR has proven to be a powerful technique for analyzing complex mixtures quickly and accurately. The first work in this field was performed at Block Engineering/Digilab using packed GC columns [8]. The work of Azarraga first showed how to interface a capillary GC to an FTIR using a "light pipe" [9]. Sensitivities in the nanogram range are routinely obtained with a light pipe GC-FTIR.

A major advance in the kinds of samples that can be analyzed by FTIR occurred with the introduction of the FTIR microscope in the early 1980s [10]. This accessory allows the infrared spectra of samples as small as 10 microns in diameter to be obtained. Both Digilab and Spectra-Tech had early products in this field. Subsequently, almost every FTIR manufacturer has offered an infrared microscope as an accessory.

The current status of FTIR technology is very exciting. A new GC-FTIR interface using cryogenic trapping [11] has pushed the minimum amount of material an FTIR can detect to less than 100 picograms. Fourier Transform Raman spectroscopy [12] promises to make this kind of vibrational analysis routine. Development of dynamic alignment and step scanning interferometers allows microsecond kinetic studies, photoacoustic depth profiling, and two dimensional infrared spectroscopy to be performed. Finally, the number of applications of the more established techniques is constantly growing, making FTIR more and more useful.

FTIRs have moved from the research lab, to the quality control lab, and on to the factory floor for on-line analysis. On the low cost end, the $10,000 barrier for a complete system may yet be broken, bringing this instrumentation into every industrial and academic lab in the country. Miniaturization and ruggedization of instruments has made portable FTIRs possible, enabling environmental sensing and remote monitoring to be done routinely. For instance, a recent article in the widely read newspaper *USA Today* discussed how a portable FTIR is being used to measure the levels of pollutants coming out of automobile tailpipes as the cars drive by the instrument [13]. A video camera takes pictures of the car's license plate, and if the vehicle's pollution levels are found to be too high, the owner of the car can be tracked down using the license plate information and given a ticket!

Interfacing an FTIR to a high pressure liquid chromatograph (HPLC) has finally been accomplished. Availability of HPLC-FTIR interfaces will greatly extend the applications and capabilities of both techniques. Obtaining the infrared spectra of biological samples is a new but rapidly expanding field. Infrared spectral differences between healthy and cancerous human colon and cervical cells have been discovered [14]. The infrared spectra of these cells have also been used to help determine the biochemical difference between healthy and malignant cells, and to help model the chemical differences between the DNA in normal and malignant tissue. If the diagnosis of cancer using FTIR is perfected, every hospital and doctor's office may eventually have one of these instruments. In the future, it will be in the environmental and biological fields where the most explosive growth of new FTIR applications will take place. The general trend will be to apply FTIR to more and more problems that impact our everyday lives. Through reducing air pollution and enhancing medical diagnoses, FTIR instruments may very well save lives and make our environment cleaner. The hundreds of scientists and engineers who have contributed to the steady improvement of FTIR instrumentation over the years should be credited with making these things possible. All told, the future looks bright for FTIR, and it is altogether fitting and proper that anyone interested in chemical analysis and identification should learn more about this technique.

C. The Advantages and Limitations of FTIR

The ultimate performance of any infrared spectrometer is determined by measuring its signal-to-noise ratio (SNR). SNR is calculated by measuring the peak height of a feature in an infrared spectrum (such as a sample absorbance peak), and ratioing it to the level of noise at some baseline point nearby in the spectrum. Noise is usually observed as random fluctuations in the spectrum above and below the baseline. An example of an isolated peak and associated baseline noise is seen in Figure 1.3. For this peak, the signal is 0.45 absorbance units at 2228 cm^{-1}, and the noise is 0.025 absorbance units at 2196 cm^{-1}. The SNR for the spectrum in this region would be (0.45/0.025) or 18.

For a given sample and set of conditions, an instrument with a high SNR will be more sensitive, be applicable to more kinds of samples, and allow

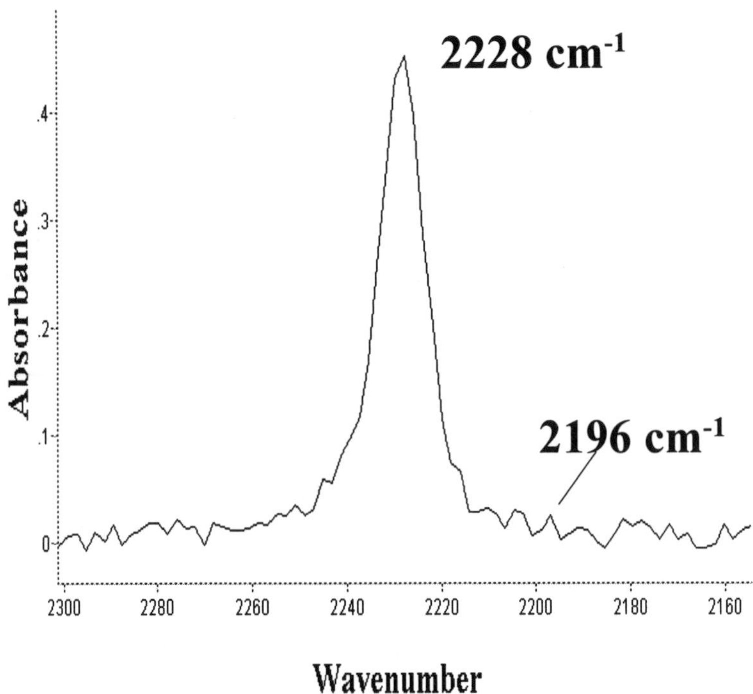

Figure 1.3 An isolated peak and associated baseline noise. The signal measured at 2228 cm^{-1} is 0.45 and the noise measured at 2196 cm^{-1} is 0.025. The SNR for this spectrum in this region is (0.45/0.025) or 18.

absorbances to be measured more accurately than an instrument with a low SNR. Signal-to noise ratio is also used to determine whether or not a spectral feature is real. A spectral feature's intensity must be 3 times that of the noise to be considered real. If a spectral feature is less than 3 times as intense as the noise, it should be ignored.

To understand the advantages of FTIR, and how it has become the predominant way of obtaining infrared spectra, the performance of FTIRs must be compared to the type of infrared instruments that came before it. These instruments were called *dispersive instruments.* An optical diagram of such an instrument is shown in Figure 1.4. Light from an infrared source passes though the sample, then goes through a slit into a *monochromator.* A monochromator contains optics which focus the light on a grating or prism to disperse the light into a spectrum of its component wavenumbers. A slit is used to select which narrow "slice" of wavenumbers is allowed to strike the detector. The use of the slit allows only a small amount of the total available light energy to strike the detector at any given time. The grating or prism is rotated so that

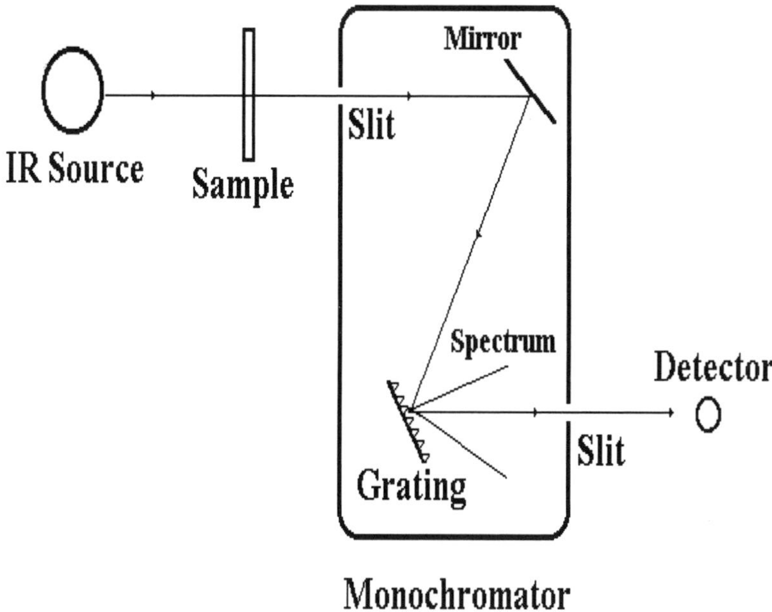

Figure 1.4 An optical diagram of a dispersive grating infrared spectrometer. The monochromator breaks the light into its component wavenumbers, then selects a narrow wavenumber range to strike the detector.

different wavenumber slices pass through the slit and are sampled by the detector. The intensity is recorded as a function of wavenumber, and plotted to produce an infrared spectrum.

There are two reasons why FTIRs are capable of SNRs significantly higher than dispersive instruments. The first is called the throughput (or Jacquinot) advantage of FTIR. It is based on the fact that all the infrared radiation passes through the sample and strikes the detector at once in an FTIR spectrometer. There are no slits to restrict the wavenumber range and reduce the intensity of infrared radiation that strikes the detector. Thus, the detector sees the maximum amount of light at all points during a scan. The second SNR advantage of FTIR is called the multiplex (or Fellgett) advantage. It is based on the fact that in an FTIR all the wavenumbers of light are detected at once, whereas in dispersive spectrometers only a small wavenumber range at a time is measured. The noise at a specific wavenumber is proportional to the square root of the time spent observing that wavenumber (this point will be discussed in more detail in Chapter 2). Because of the multiplex advantage, acquiring data for 10 minutes on an FTIR means all wavenumbers are observed for a full 10

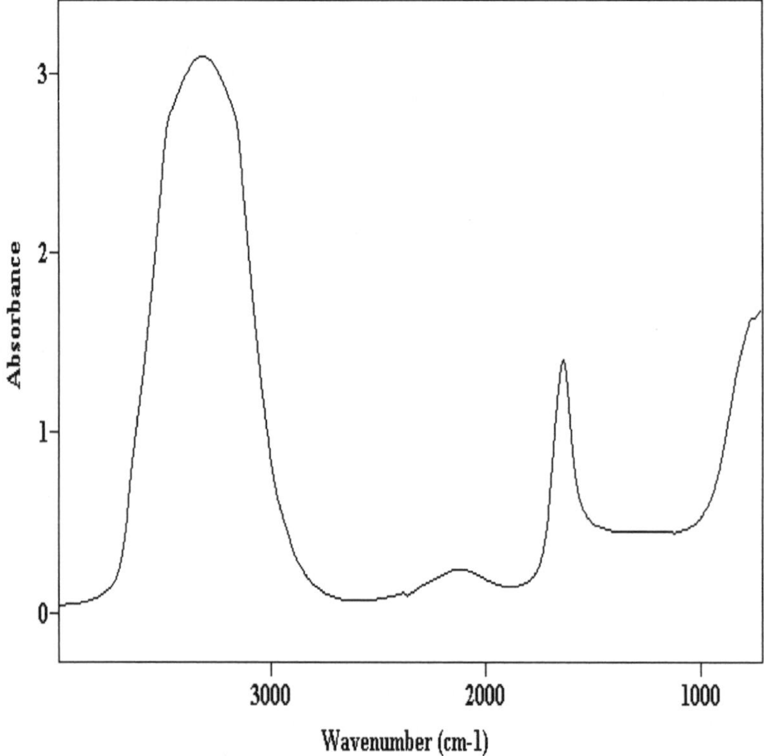

Figure 1.5 The infrared spectrum of liquid water, obtained using the liquid ATR technique. Note the intensity and width of the absorbance bands.

minutes. In a dispersive spectrometer each wavenumber is observed for only a small fraction of the 10 minute measurement time.

The practical advantage of multiplexing is that an FTIR can acquire a spectrum much faster than a dispersive instrument. This allows multiple scans of the same sample to be added together. For a constant resolution, the SNR is related to the number of scans added together as follows:

$$SNR \propto (N)^{1/2}$$

where N is the number of scans added together, and $\propto$ stands for proportionality. As an example, a single 4 cm^{-1} resolution scan from 4000 to 400 cm^{-1} on a dispersive instrument takes about 10 minutes. The same spectrum on an FTIR takes one second, and the FTIR can obtain 600 scans in 10 minutes. As a result, the FTIR can achieve an SNR advantage of $(600)^{1/2}$ (a factor of 24.5) versus a dispersive instrument for the same amount of analysis time. The improved SNRs available with FTIR explains why they have become the instrument of choice for obtaining infrared spectra, and also explains the large number of new infrared applications that have been developed.

Despite the many advantages of FTIR , there are limitations on what is achievable with infrared spectroscopy in general, and there is one specific limitation of FTIR in particular. A general limitation of infrared spectroscopy is that it cannot detect atoms or monatomic ions. Single atomic entities contain no chemical bonds, do not possess vibrational motion, and hence do not absorb infrared radiation. In general, it is impossible for an infrared spectrometer to measure the level of an element in a substance, unless the element is present as part of a molecule whose spectrum can be detected. Also, the noble gases such as helium and argon can not be detected because they exist as individual molecules. Another class of substances that do not absorb infrared radiation are homonuclear diatomic molecules. These are molecules comprised of two identical atoms, such as N_2 and O_2. Homonuclear diatomic molecules do not possess infrared spectra due to their symmetry. N_2 and O_2 are very abundant in the atmosphere, and if they absorbed infrared radiation all the radiation in the spectrometer would be absorbed by the atmosphere before it got to the sample.

Another limitation of infrared spectroscopy is its use in analyzing complex mixtures. These samples give rise to complex spectra, whose interpretation is difficult because it is hard to know which bands are from which molecules in a sample. Infrared spectroscopy works best on pure substances, since all bands can be assigned to a specific chemical structure. Before obtaining an infrared spectrum, it is best to separate out the components of complex mixtures using methods such as recrystallization, distillation, and chromatography. Afterwards, the spectrum of each component can be obtained separately. The author realizes that the vast majority of real world samples are complex mixtures, and that separating them into components is not always possible. However, to maximize the information obtained from an infrared spectrum, spectra of individual molecules should be obtained. Another technique that can be used on mixtures is to take the spectra of pure molecules known to be in the sample and subtract them from the mixture spectrum. This will be discussed in Chapter 3.

Aqueous solutions are also difficult to analyze using infrared spectroscopy. Water is a strong infrared absorber, as seen in Figure 1.5, and dissolves many infrared transparent materials used in cells and as windows. A technique known as liquid ATR (see Chapter 5) enables one to obtain infrared spectra of aqueous solutions. However, the water spectrum must be subtracted from the sample spectrum, and the sensitivity of the technique is in the 0.1 - 1% range. Often times the water absorbance is too strong to subtract out properly, masking regions of the sample's spectrum. The detection of small quantities of molecules in water is possible using GC-FTIR, but other techniques may give better results when analyzing aqueous solutions.

A specific limitation of FTIR spectrometers is that FTIR is a single beam technique. This means that the background spectrum, which measures the contribution of the instrument and the environment to the spectrum, is measured at a different point in time than the spectrum of the sample. Ideally, the instrumental and environmental contributions to the spectrum are eliminated by ratioing the sample spectrum to the background spectrum. However, if something in the instrument or the environment changes between when the sample and background spectra are obtained, spectral artifacts can appear in the sample spectrum. Common examples include water vapor and carbon dioxide peaks. These artifacts can be misinterpreted, or may mask sample absorbances. Artifacts do not always appear in sample spectra, but one should know where they may show up to avoid misassigning them. Older style dispersive instruments were double beam instruments, which means the environmental and instrumental contributions to the spectrum were compensated for in real time during the measurement of an infrared spectrum. Artifacts such as water vapor peaks do not appear in spectra taken on dispersive instruments. This is the one advantage that dispersive instruments have over Fourier transform instruments.

Infrared spectroscopy is a wonderful tool for detecting functional groups, but by itself cannot necessarily be used to elucidate the complete structure of an unknown molecule. Often times an infrared spectrum does not contain enough information, or contains misleading or contradictory information, making the complete determination of an unknown's structure impossible. The best approach to this problem is to use FTIR in conjunction with other molecular spectroscopy techniques such as NMR, mass spectrometry, UV/VIS spectroscopy, and Raman scattering. All these techniques provide different pieces of information about a molecule's structure, and together provide a powerful and effective means of identifying unknowns. Thus, one must put the use of FTIR in perspective. It is a very useful analytical technique capable of solving many problems, but it is not the answer to every chemical analysis problem.

For your convenience, the names and addresses of the major companies manufacturing and selling FTIRs in the United States has been included below.

Names and Addresses of Major FTIR Company Offices in the U.S.A.

Bomem Division of Applied Automation
7780 Quincy St.
Willowbrook, IL 60521
(800) 858-FTIR

Mattson Instruments
1001 Fourier Dr.
Madison, WI 53717
(608) 831-5515

Bruker Instruments
Manning Park
Billerica, MA 01821
(508) 667-9580

Digilab Division of Bio-Rad
237 Putnam Ave.
Cambridge, MA 02139
(617) 868-4330

KVB/Analect
17819 Gillete Ave.
Irvine, CA 92714
(714) 660-8801

Midac Corporation
17911 Fitch Ave.
Irvine, CA 92714
(714) 660-8558

Nicolet Instrument Corp.
5225 Verona Rd.
Madison, WI 53711
(608) 276-6100

Perkin-Elmer Corp.
761 Main Ave.
Norwalk, CT 06859
(203) 762-4000

References

1. Dorothy Michelson Livingston, *The Master of Light: A Biography of Albert Abraham Michelson*, The University Press of Chicago, Chicago, 1973.
2. A. Michelson, *Phil. Mag.* **31**(1891)338, **34**(1892)280.
3. J.W. Cooley and J.W. Tukey, *Math. Comput.* **19**(1965)297.
4. L. Mertz, *Astron. J.* **70**(1965)548.
5. P.R. Griffiths, R. Curbelo, C.T. Foskett, and S.T. Dunn, *Analytical Instrumentation (Inst. Society of America)* **8**(1970)II-4.
6. N.J. Harrick, *Internal Reflection Spectroscopy*, Wiley, New York, 1967.
7 M. Fuller and P. Griffiths, *Anal. Chem.* **50**(1978)1906.
8. M. Low and S. Freeman, *Anal. Chem.* **39**(1967)194.
9. L.V. Azarraga, *Appl. Spec.* **34**(1980)224.
10. R. Messerchmidt and M. Harthcock, Eds. *Infrared Microspectroscopy: Theory and Applications*, Marcel Dekker, New York, 1988.
11. A. Haefner, K. Norton, P. Griffiths, S. Bourne, and R. Curbelo, *Anal. Chem.* **60**(2441)1988.
12. T. Hirschfeld and B. Chase, *Appl. Spec.* **40**(1986)133.
13. *USA Today* June 23, 1993, pg. 8A.
14. P. Wong, S. Lacele, H. Yazdi, *Appl. Spec.* **47**(1993)1830.

Bibliography

S. Johnston, *Fourier Transform Infrared: A Constantly Evolving Technology*, Ellis Horwood, London, 1991.

Chapter 2

How an FTIR Works

A. How an Interferometer Works

The purpose of an interferometer is to take a beam of light, split it into two beams, and make one of the light beams travel a different distance than the other. The difference in distance travelled by these two light beams is called the *optical path difference* (or optical retardation), denoted by the Greek letter δ (small delta). In this chapter we will restrict our discussion to the Michelson interferometer, since it was the first interferometer to be used in commercial FTIR instruments, and is still at the heart of most FTIRs in use. Other interferometer designs have been invented and commercialized, but the principles developed here in discussing the Michelson interferometer are applicable to all interferometer designs. A diagram of a Michelson interferometer is shown in Figure 2.1. The Michelson interferometer consists of four arms. The first arm contains a source of infrared light, the second arm contains a stationary (fixed) mirror, the third arm contains a moving mirror, and the fourth arm is open. At the intersection of the four arms is a beamsplitter, which is designed to transmit half the radiation that impinges upon it, and reflect half of it. As a result, the light transmitted by the beamsplitter strikes the fixed mirror, and the light reflected by the beamsplitter strikes the moving mirror. After reflecting off their respective mirrors, the two light beams recombine at the beamsplitter, then leave the interferometer to interact with the sample and strike the detector.

If the moving mirror and fixed mirror are the same distance from the beamsplitter, the distance travelled by the light beams that reflect off these mirrors is the same. This condition is known as *zero path difference* (or ZPD). In a Michelson interferometer an optical path difference is introduced between the two light beams by translating the moving mirror away from the beamsplitter. The light that reflects off the moving mirror will travel further than the light that reflects off the fixed mirror. The distance that the mirror is moved from ZPD is called the *mirror displacement,* and is denoted by the

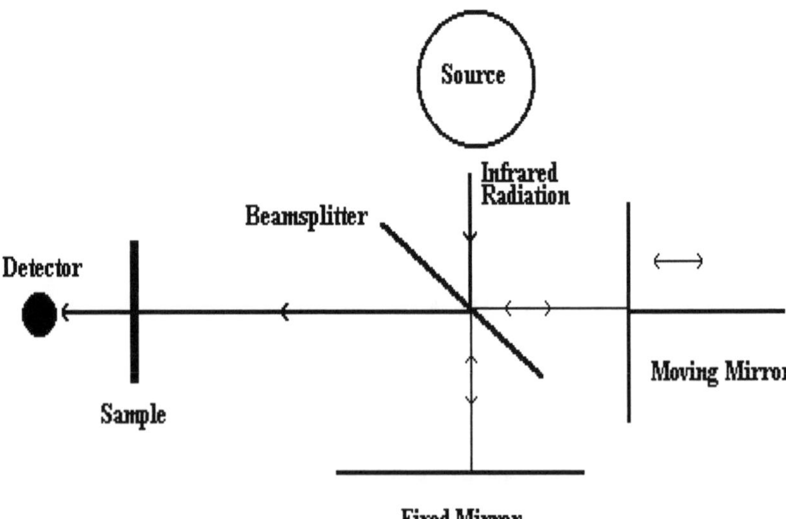

Figure 2.1 A optical diagram of a Michelson interferometer.

symbol Δ. Imagine for a moment that the mirror displacement has some value Δ. The light that reflects off the moving mirror travels a distance Δ on the way to the mirror and a distance Δ on its return trip to the beamsplitter. The extra distance this light beam travels compared to the fixed mirror light beam is equal to 2Δ. Thus, the relationship between mirror displacement and optical path difference is

$$\delta = 2\Delta$$

Imagine using a monochromatic light source, such as a laser (although lasers are not the light source used in an FTIR) of wavelength λ. Also, assume the interferometer is at ZPD as shown in Figure 2.1. When the beams that have reflected off the fixed and moving mirrors recombine at the beamsplitter, they will be in phase. Their crests and troughs will overlap as shown in Figure 2.2. The beams leave the beamsplitter in phase, and since they travel the same distance at the same speed, they are in phase when they recombine at the beamsplitter. Now, it is a general property of waves that their amplitudes are additive. Two waves colliding on the ocean to produce a more intense wave is an example of this phenomenon. The amplitudes of the two light beams shown in Figure 2.2 add to give a light beam whose amplitude is greater than the amplitude of each of the individual beams. This phenomenon is known as *constructive interference*. As a result of constructive interference, an intense light beam leaves the interferometer. All wavelengths of light constructively interfere at ZPD. Constructive interference also takes place when the

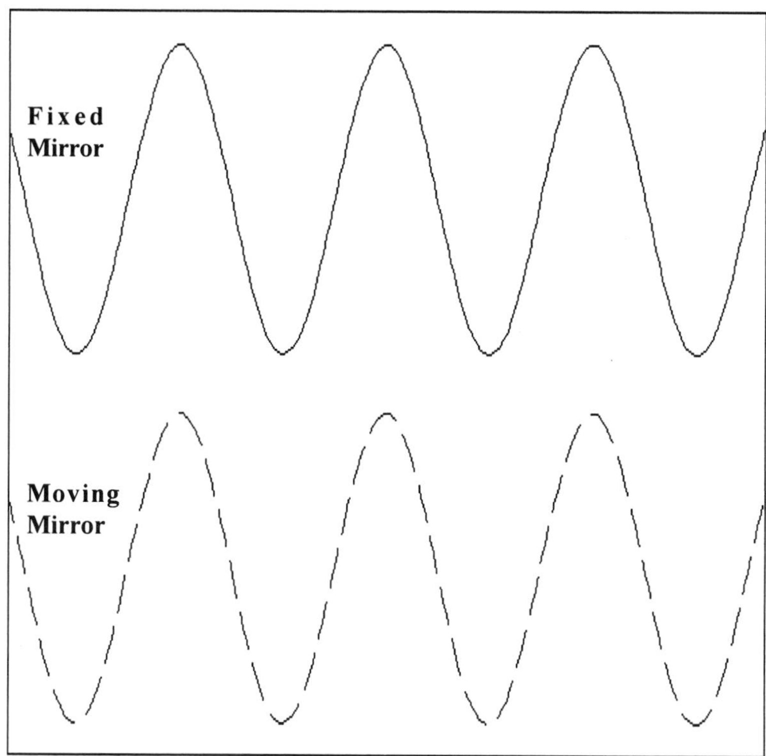

Figure 2.2 An illustration of the fixed mirror light beam (solid line) and moving mirror light beam (dashed line) in a Michelson interferometer when the optical path difference is zero (ZPD). The two beams are completely in phase and constructively interfere, combining to give an intense light beam leaving the interferometer.

optical path difference is equal to multiples of λ. In this case, the light beams have travelled different distances, but are out of phase the right amount to lead to a perfect overlap of crests with crests (and troughs with troughs). Constructive interference will take place for any value of δ where the two light beams are in phase. This is summarized in equation form as follows

$$\delta = n\lambda \qquad (2.1)$$

where n is any integer with the values n = 0,1,2,3... . ZPD corresponds to when n = 0, and totally constructive interference will take place whenever δ is some multiple of the optical path difference.

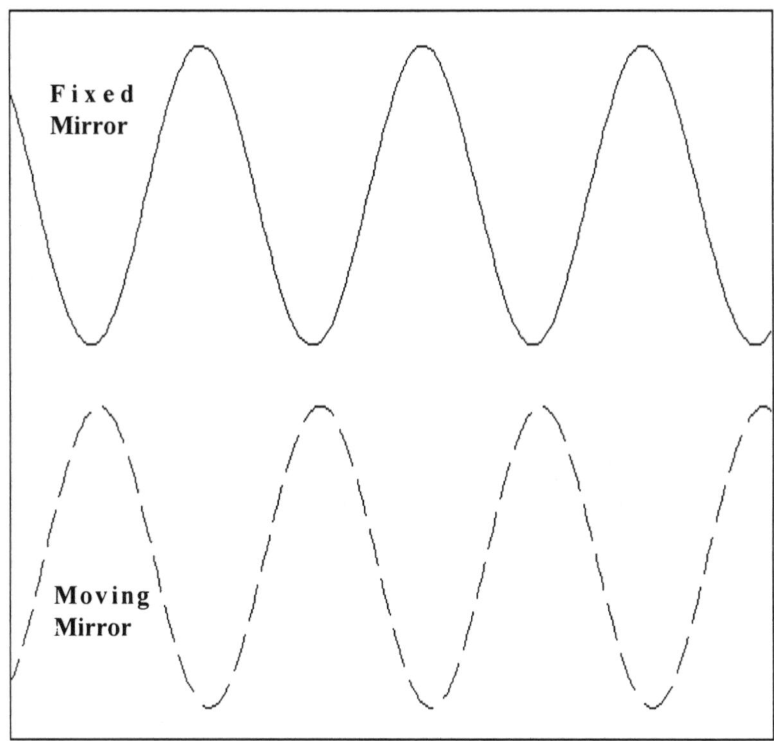

Figure 2.3 An illustration of the fixed mirror light beam (solid line) and the moving mirror light beam (dashed line) when the optical path difference between the two beams is 1/2 of a wavelength of light ($\delta = 1/2\lambda$). The two beams are completely out of phase and destructively interfere, combining to give a weak light beam leaving the interferometer.

Imagine changing the mirror displacement of a Michelson interferometer so that the moving mirror is $1/4\lambda$ away from ZPD. Since $\delta = 2\Delta$, the optical path difference is $1/2\lambda$. When the fixed mirror and moving mirror light beams are recombined at the beamsplitter they are completely out of phase, as seen in Figure 2.3. The crests and troughs of the light waves overlap, and their amplitudes add together and cancel each other, giving a light beam of low intensity. This is known as *destructive interference*. It is analogous to the crest of an ocean wave encountering a trough; the two cancel to give a calm sea. Totally destructive interference takes place when the optical path difference is $1/2\lambda$ or some multiple of it. In this case the light beams are out of phase the right amount to lead to a perfect overlap of crests with troughs. The

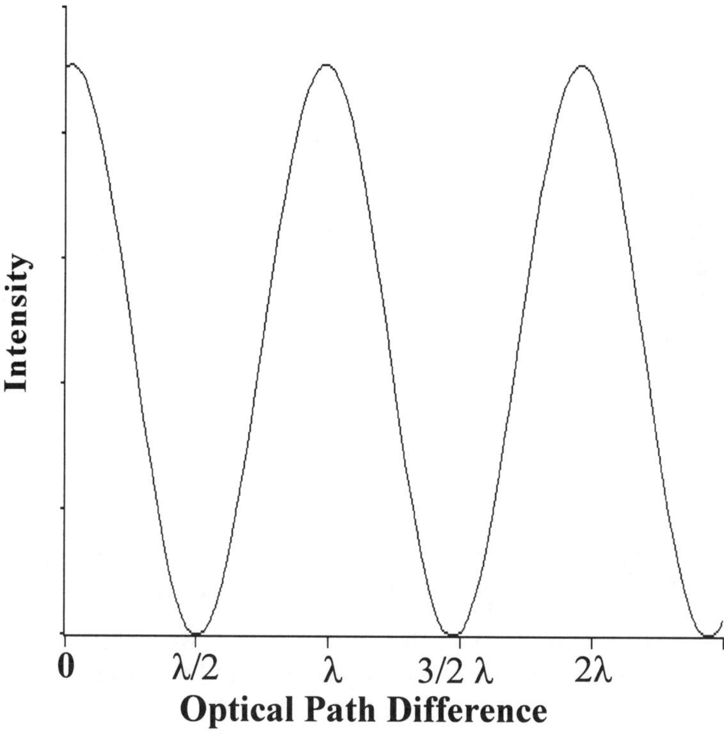

Figure 2.4 The interferogram of a monochromatic light source of wavelength λ.

optical path differences at which totally destructive interference takes place is given by:

$$\delta = (n + 1/2)\lambda \qquad (2.2)$$

where n = 0,1,2,3... .

At optical path differences other than those given in equations 2.1 and 2.2, a combination of constructive and destructive interference takes place, and the light beam intensity is somewhere between very bright and very weak. If the mirror is moved at constant velocity, the intensity of the infrared radiation increases and decreases smoothly. The variation of light intensity with optical path difference is measured by the detector as a sinusoidal wave (more specifically a cosine wave). A plot of light intensity versus optical path difference is called an *interferogram*. The interferogram of light of wavelength

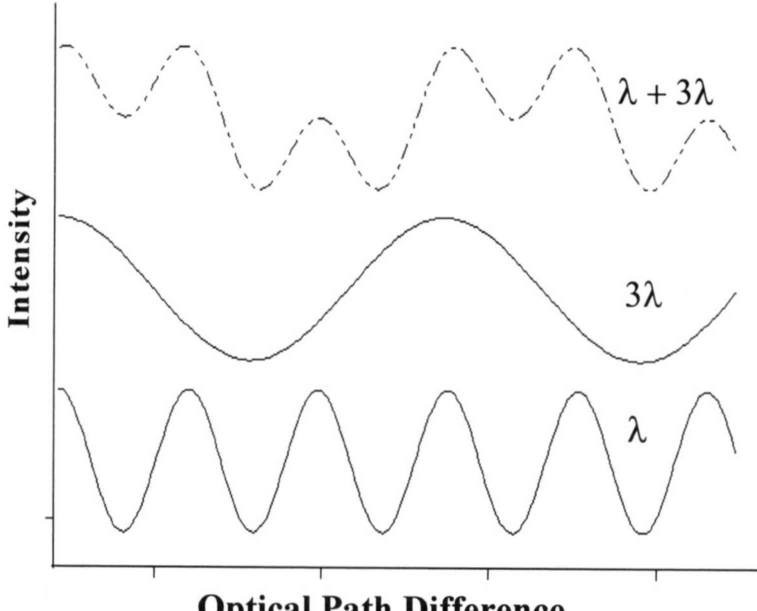

Figure 2.5 Bottom: The interferogram of light of wavelength λ. Middle: The interferogram of light of wavelength 3λ. Top: The sum of the two interferograms.

λ is shown in Figure 2.4. The fundamental measurement obtained by an FTIR is an interferogram, which is Fourier transformed to give a spectrum. This is where the term *Fourier transform infrared spectroscopy* comes from. To generate a complete interferogram, the moving mirror is translated back and forth once. This is known as a *scan.*

The constructive and destructive interference that takes place in the interferometer affects the light intensity at a given wavelength as if a shutter were opening and closing in the light beam, alternately blocking the beam and letting light through. Therefore, a light beam that passes through an interferometer is said to be modulated. Modulated light beams are denoted by the number of times per second they switch between being light and dark, known as their frequency. The frequency at which light passing through an interferometer is modulated is given by the following equation:

$$F_v = 2VW \qquad (2.3)$$

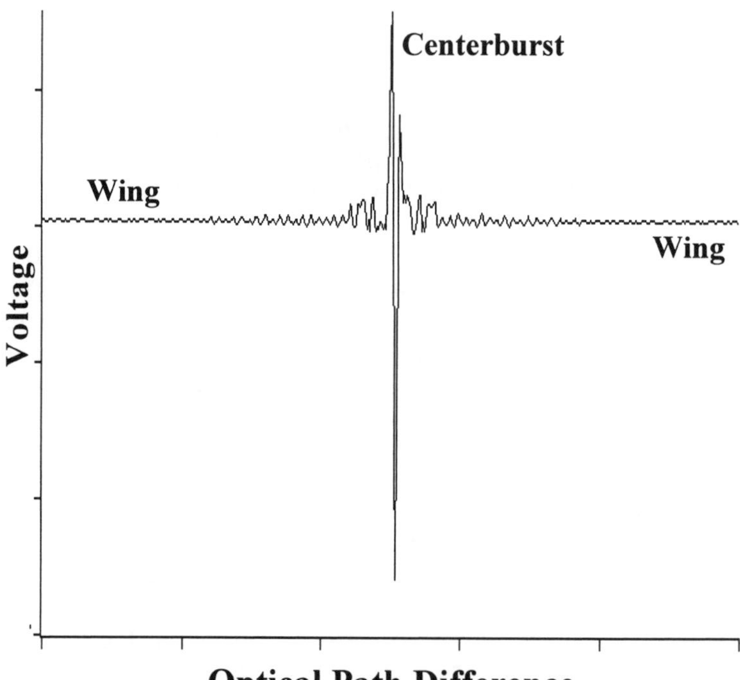

Figure 2.6 The interferogram of a broadband infrared source. Note the centerburst, which corresponds to ZPD, and the wings. The Y axis is in voltage since this is the unit in which the detector measures infrared intensity.

where F_v is the modulation frequency, V is the moving mirror velocity in cm/sec, and W is the wavenumber of the light in the interferometer measured in cm^{-1}. Equation 2.3 gives the frequency of the cosine wave interferogram that is measured by the detector for light of wavenumber W passing through the interferometer. For repeated measurements on the same sample to be reproducible, the interferogram, and hence F_v, must be reproducible. This is why it is critical that the velocity of the moving mirror in an FTIR be closely controlled.

Imagine light of wavelengths λ and 3λ passing through the interferometer shown in Figure 2.1. By substituting 3λ for λ in equations 2.1 and 2.2, we find light of this wavelength undergoes constructive interference when

$$\delta = 3n\lambda$$

and destructive interference occurs when

$$\delta = 3(n + 1/2)\lambda$$

Light of wavelengths λ and 3λ undergo totally constructive and destructive interference at different optical path differences. As a result, the interferograms of the two wavelengths will be different. A plot of the interferograms for light of wavelengths λ and 3λ is shown in the middle and bottom of Figure 2.5. Both of these interferograms are cosine waves, but their frequencies are different as expected from equation 2.3. This is how the signals due to different wavenumbers of light are distinguished by the spectrometer. Each different wavenumber of light gives rise to a sinusoidal wave signal of unique frequency that is measured by the detector. Interferogram signals are additive, so the actual signal measured by the detector will be the sum of the two cosine waves seen in the top of Figure 2.5.

Imagine replacing the monochromatic light source in the interferometer with a broadband infrared source (as is the case in FTIR). These sources give off light at a continuum of wavelengths, each one of which gives rise to a different cosine wave interferogram whose frequency is given by equation 2.3. The total interferogram measured by the detector is the summation of all the interferograms from all the different infrared wavelengths. In fact, the measured intensity of light at a given λ is determined by the amplitude of that wavelength's interferogram. The interferogram of a real world broadband infrared source is shown in Figure 2.6. Since the intensity information for all the different wavelengths of light are contained in one measured interferogram, there is no need to physically separate the light beam into its component wavelengths and measure intensities one at a time as in dispersive IR spectroscopy (see Chapter 1). In essence, a Fourier transform is performed on the light spectrum as it passes through the interferometer. This generates the interferogram, which must be Fourier transformed to get the spectrum back. A simplistic way to think about the process is that the interferometer "encodes" the intensity and wavelength information so all the data can be measured at once, and the Fourier transform decodes the information to obtain the spectrum.

The interferogram in Figure 2.6 displays a sharp intensity spike at zero path difference, called the *centerburst*, and low intensity at higher optical path differences in the *wings* of the interferogram. The high intensity of the centerburst is caused by all wavelengths constructively interfering at zero path difference. All the component sinusoidal waves that comprise the interferogram have ZPD as a common point. As optical path difference is increased, sinusoidal waves of different frequency get out of phase with each other and destructively interfere. As a result, the interferogram's intensity drops off rapidly as the mirror moves away from ZPD and into the wings.

Once light leaves the interferometer it passes through the sample compartment and is focused upon the detector. The detector is simply an electrical transducer, and produces an electrical signal, be it voltage, resistance, or current, that is proportional to the amount of infrared radiation striking it. Thus,

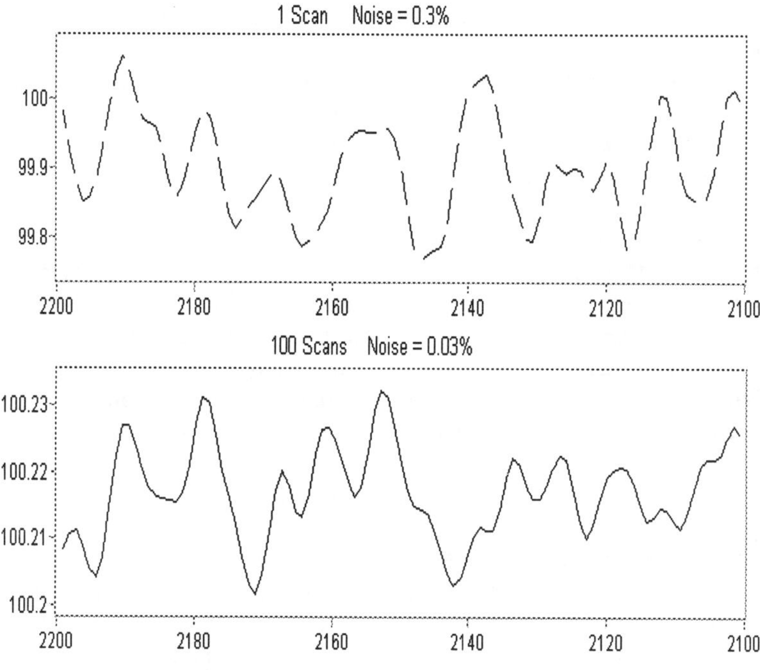

Figure 2.7 Top: A spectrum with a noise level of 0.3% after one scan. Bottom: A spectrum obtained under the same conditions by coadding 100 scans. The noise level is 0.03%, a 10-fold improvement in SNR as expected from eq. 2.4.

Table 2.1
The Relationship between Number of Scans and Signal-to-Noise Ratio

# of Scans	SNR Improvement over 1 Scan
1	-
4	2
16	4
64	8
100	10
256	16
1000	31.6

an interferogram is truly a plot of detector response versus optical path difference.

By adding interferograms together in an FTIR, random noise is reduced and the signal-to-noise ratio is improved as pointed out in Chapter 1. This process is known as *coadding.* Here is how coadding improves SNR. From scan to scan, the presence of the signal in an interferogram due to an absorbance band is constant, but the magnitude and sign of noise is random. For simplicity, assume we are adding infrared spectra together instead of interferograms. Lets also assume that during scan 1 there is an absorbance band of 0.5 absorbance units (A.U.) at 2000 cm^{-1}, and the noise at 2150 cm^{-1} is +0.04 A.U. For scan 2 the absorbance at 2000 cm^{-1} will still be 0.5 (unless the sample concentration has changed somehow between scans), but since the sign of noise is random, the noise value at 2150 cm^{-1} may be -0.02 A.U. When scans 1 and 2 are added together, the noise level at 2150 cm^{-1} is reduced as follows:

$$(0.04) + (-0.02) = 0.02$$

The positive and negative fluctuations in the random noise level cancel themselves out as more scans are added together. This is how the multiplex advantage of FTIR (discussed in Chapter 1) works. The relationship between signal-to-noise ratio (SNR) and the number of scans obtained at a given resolution is

$$\text{SNR} \propto (\text{N})^{1/2} \qquad (2.4)$$

where N is the number of scans added together at a given resolution. Thus, a spectrum consisting of 100 coadded spectra would have an SNR 10 times better than a spectrum comprised of just one scan ($(100)^{1/2} = 10$). This affect is shown in Figure 2.7. Table 2.1 lists the SNR improvement for various numbers of scans compared to a single scan.

According to equation 2.4, it appears that scans can be coadded indefinitely until a specified SNR is achieved. This isn't true in practice. During long data collects, drifts in instrument response (amongst other factors) can contribute noise to the spectrum. Since the sign of this kind of noise is not random, it will not cancel out when scans are added together. As a result, there is a limit beyond which increasing the number of scans does not improve SNR. This limit depends on the nature of the specific FTIR being used.

A discussion of how many scans to use for a given sample is in order. Generally, for routine samples analyzed in the instrument's sample compartment, coadding 100 scans or less is usually sufficient to obtain a reasonable SNR. Occasionally, difficult to analyze samples, or samples analyzed outside the sample compartment using techniques such as infrared microscopy, may require coadding upwards of 1,000 scans to achieve a usable SNR. Coadding more than 1,000 scans to obtain an FTIR spectrum is not recommended since

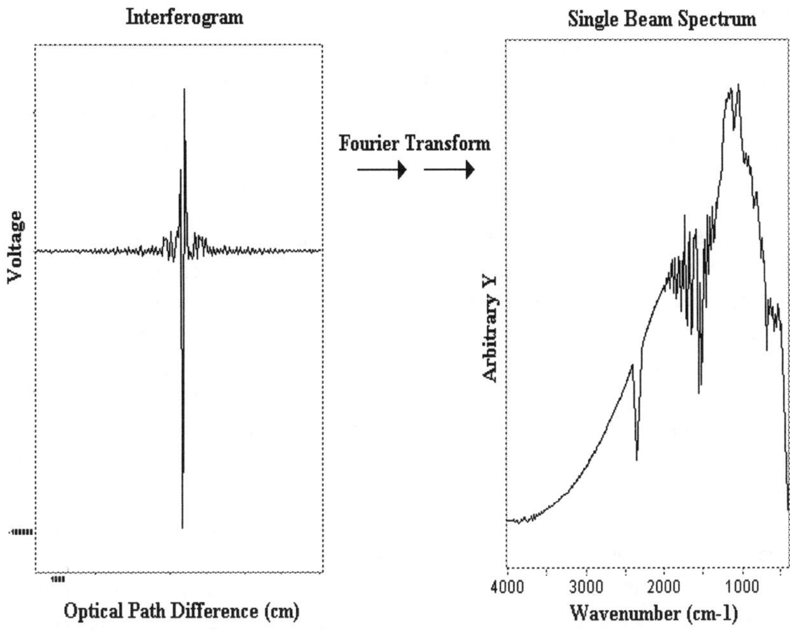

Figure 2.8 An illustration of how an interferogram is Fourier transformed to generate a single beam infrared spectrum.

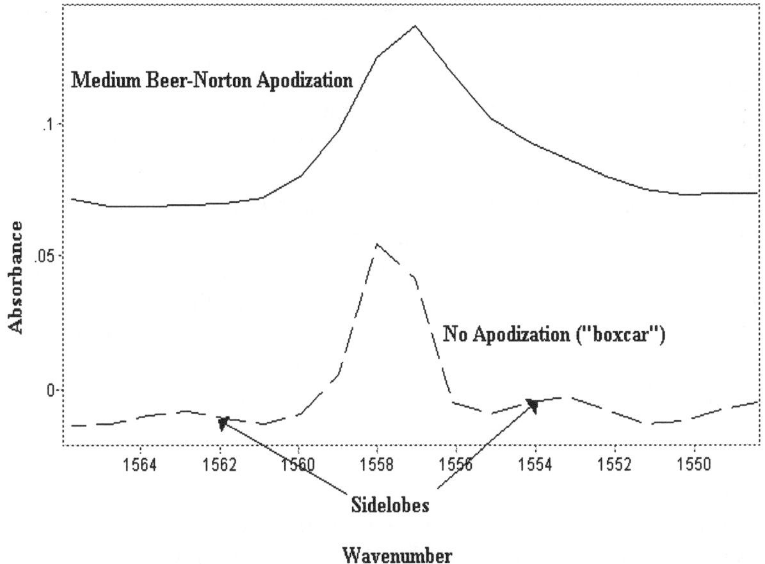

Figure 2.9 The spectrum of water vapor with boxcar apodization (bottom), and with a "medium Beer-Norton" apodization function (top).

it takes a long time, and as mentioned above, may not result in an improvement in SNR. If coadding 1,000 scans does not produce a good spectrum, try concentrating the sample, using a lower resolution to increase SNR, or use a different sampling technique. If all else fails, perhaps the use of analytical techniques other than FTIR should be considered for your sample.

B. How an Interferogram Becomes a Spectrum

As stated above, an interferogram is simply a large number of sinusoidal signals added together. According to a theorem developed by Fourier, any mathematical function (an x, y plot) can be expressed as a sum of sinusoidal waves. For the moment, think of an infrared spectrum as a mathematical function. An interferogram is a sum of sinusoidal waves, each of which contains information about the wavenumber of a given infrared peak and amplitude information about the peak intensity at that wavenumber. The Fourier transform simply calculates the infrared spectrum from the summed sinusoidal waves in the interferogram.

There is another way to think about the Fourier transform that, although it is not rigorously correct, is useful in illustrating how the Fourier transform works. The interferogram is a plot of infrared intensity versus optical path difference, which can be measured in cm. By its very nature, when a function is Fourier transformed its X axis units are inverted. The Fourier transform of an interferogram produces a plot of infrared intensity versus inverse centimeters, or cm^{-1}, as shown in Figure 2.8. Inverse centimeters are also known as wavenumbers. A plot of infrared intensity versus wavenumber is an infrared spectrum. In a sense, the Fourier transform inverts the interferogram to produce the infrared spectrum.

Calculating a Fourier transform involves performing a mathematical integral on the interferogram. Ideally, the limits of this integral should be plus and minus infinity. This means one must use an optical pathlength of infinity, and collect an infinite number of data points to obtain an interferogram that will Fourier transform properly. This is impossible, so the interferogram and the integral must be truncated at some finite point, and the limits of integration are ZPD and the maximum optical path difference measured. An unfortunate outcome of truncating the integral is that the lineshapes of the peaks in the infrared spectrum are affected. Rather than obtaining the true peak shape, the absorbance bands are surrounded by sidelobes, which are sinusoidal undulations in the baseline. Sidelobes are illustrated in the bottom of Figure 2.9. These lobes, or “feet” as they are called, are suppressed by multiplying the interferogram by an *apodization function*. The word apodization comes from the Greek words “a podi” (α ποδι) or “no feet”. The interferogram is Fourier transformed after this multiplication. A side effect of using apodization functions is that spectral resolution is reduced. An example of this effect is shown in Figure 2.9. The bottom spectrum was obtained with boxcar apodization, and shows sidelobes. The top spectrum was obtained by transforming the same interferogram using a "medium Beer-Norton" apodization function. This

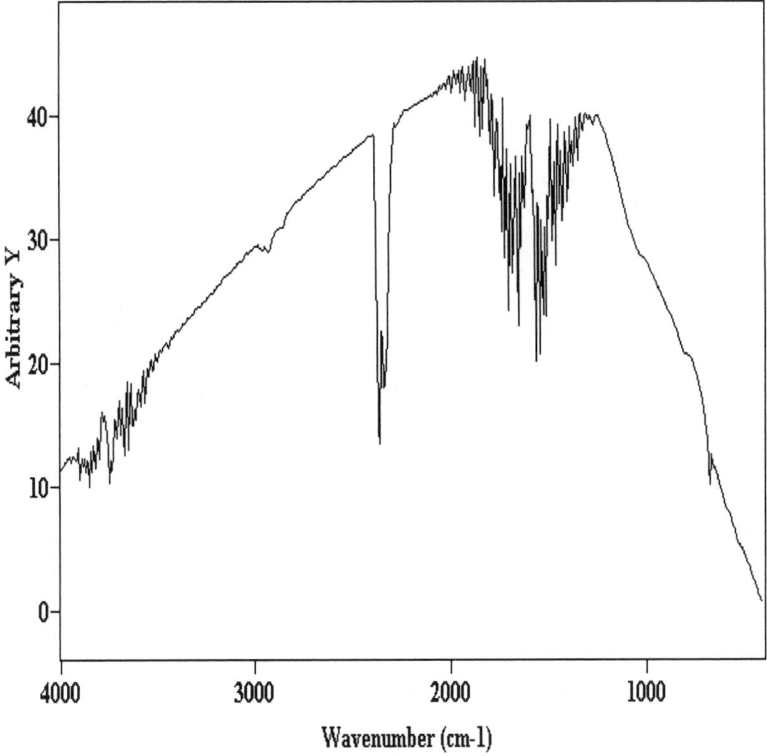

Figure 2.10 A single beam background spectrum.

spectrum's baseline has no sidelobes, but the peak is noticeably broadened compared to the bottom spectrum. The apodization function that provides the highest resolution, and does the worst job of suppressing sidelobes, is the boxcar apodization function. This function should only be used when high resolution is of the utmost importance, and the user understands and can live with the existence of side lobes. Apodization functions vary in how well they suppress sidelobes, and how much they degrade resolution. The apodization function is a user selectable software parameter in most manufacturers' FTIR systems, and the types of functions available differ from instrument to instrument. For quantitative analysis, the "medium Beer-Norton" function gives the best results [1]. It is up to the user to experiment with apodization functions, and determine which one works best for a given application.

When an interferogram is Fourier transformed, a *single beam spectrum* is obtained as shown in Figure 2.10. A single beam spectrum is a plot of raw detector response versus wavenumber. A single beam spectrum obtained without a sample in the infrared beam is called a *background spectrum*. The back-

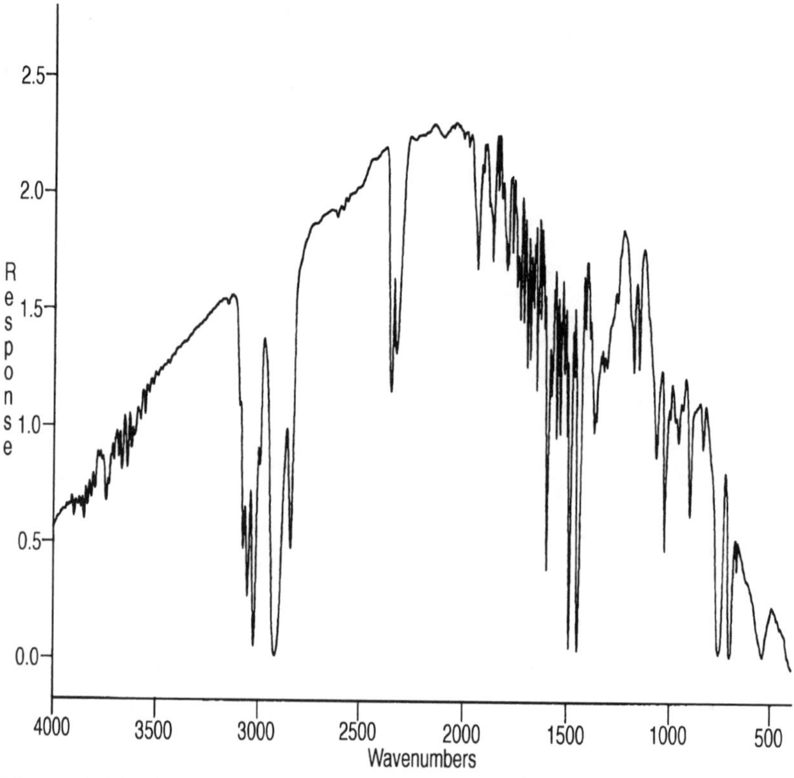

Figure 2.11 The sample single beam spectrum of polystyrene.

ground spectrum contains the instrument's and the environment's contribution to the infrared spectrum. The contribution of the instrument by itself to the background spectrum is known as the *instrument response function.* The spectrum seen in Figure 2.10 is a background spectrum. The overall shape of the spectrum is due to the sensitivity of the detector, transmission and reflection properties of the beamsplitter, emissive properties of the source, and reflective properties of the mirrors. Common features around 3500 and 1630 cm^{-1} are due to atmospheric water vapor, and the bands at 2350 and 667 are due to carbon dioxide. A background spectrum must always be run when analyzing samples by FTIR.

When an interferogram is measured with a sample in the infrared beam and Fourier transformed, a sample single beam spectrum is produced as shown in Figure 2.11. It looks similar to the background spectrum except the sample peaks (in this case polystyrene) are superimposed upon the instrumental and atmospheric contributions to the spectrum. To eliminate these contributions, the sample single beam spectrum must be ratioed against the background spec-

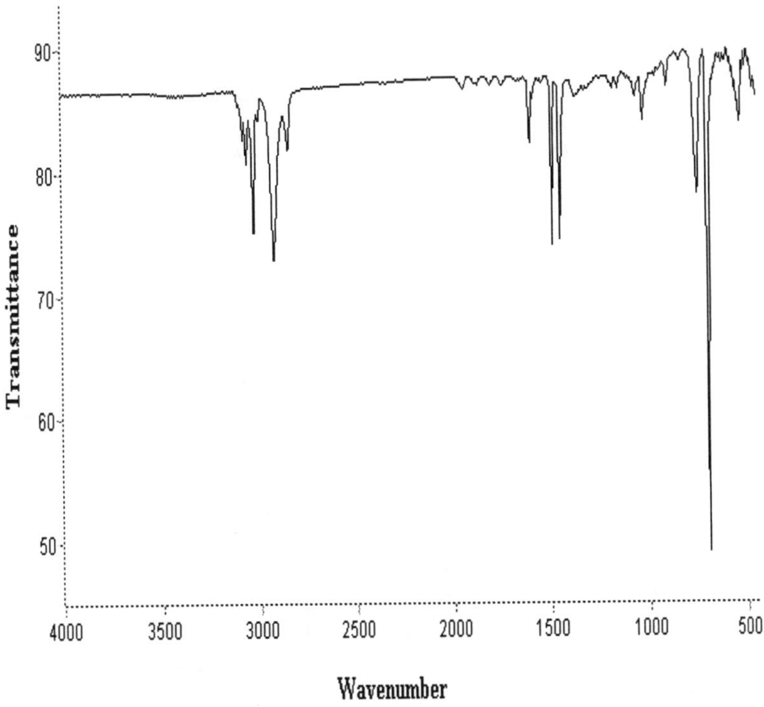

Figure 2.12 The transmittance spectrum of polystyrene.

trum. This produces a transmittance spectrum as shown by the following equation.

$$\%T = I/I_0$$

where:

$\%T$ = Transmittance
I = Intensity measured with a sample in the beam (from the sample single beam spectrum)
I_0 = Intensity measured with no sample in the beam (from the back ground spectrum)

The absorbance spectrum can be calculated from the transmittance spectrum using the following equation.

$$A = -\log_{10} T$$

where:

Table 2.2
List of Absorbance Wavenumbers of Atmospheric Gases

Gas	Wavenumbers
CO_2	2350, 667
H_2O	3900-3400, 1850-1350

T = Transmittance
A = Absorbance

Ideally, the final spectrum should be devoid of all instrumental and environmental contributions, and only contain features due to the sample. Figure 2.12 shows a quality transmittance spectrum of a polystyrene film, generated by ratioing Figure 2.11 to Figure 2.10. For qualitative analysis, the choice of whether to use spectra in absorbance or transmittance is up to the user. However, absorbance units must be used for quantitative analysis.

As mentioned above, water vapor and carbon dioxide absorb in the mid-infrared, and their bands are easily seen in single beam spectra. If the concentrations of these gases in the instrument are the same when the background and sample spectra are obtained, their contributions to the spectrum will ratio out exactly and their bands will not be seen. If the concentrations of these gases are different when the background and sample spectra are obtained, their bands will be seen in the sample spectrum. If working in transmittance, these peaks will point down if the gas concentrations are greater when the sample is run compared to when the background spectrum was run. The gas bands will point up if their concentrations are greater when the background is run than when the sample is run (for absorbance spectra, the gas bands would point up and down respectively).

Purging FTIR spectrometers with dry nitrogen reduces the concentration of H_2O and CO_2 to some extent, but it can be very difficult to reproduce CO_2 and H_2O levels in the instrument when opening and closing the sample compartment to change samples. To try and reproduce the gas concentrations when the background and sample spectra are run, try the following trick. Open and close the sample compartment, wait a minute, then run a background spectrum. Open the sample compartment and place the sample in the infrared beam, holding the sample compartment open for the same amount of time as before the background spectrum. Close the sample compartment, wait one minute, then obtain the sample spectrum. This will hopefully reproduce the concentrations of CO_2 and H_2O in the spectrometer so that their bands ratio out. Despite your best efforts, many times there are some residual water vapor and carbon dioxide bands in FTIR spectra. To avoid misinterpreting CO_2 and H_2O bands, their major absorbances are listed in Table 2.2.

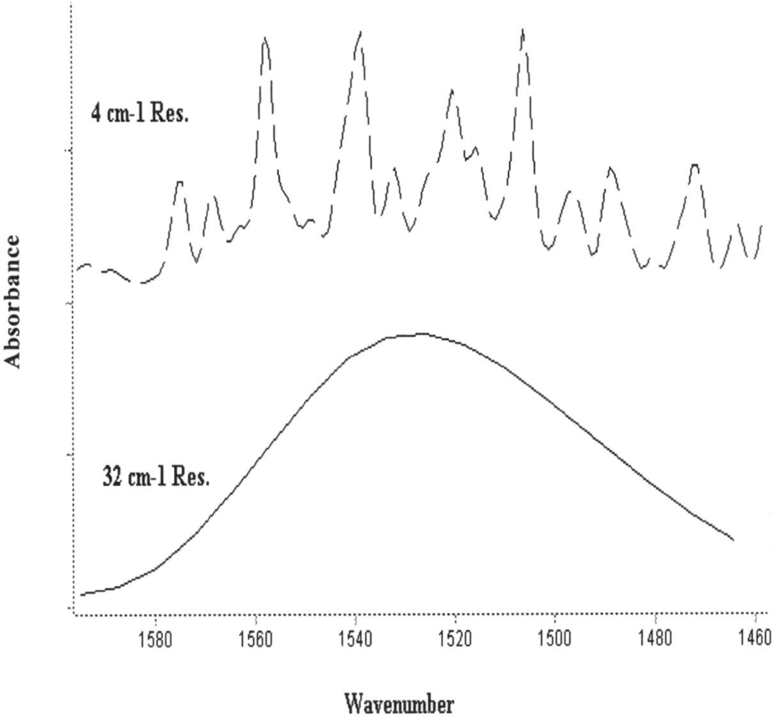

Figure 2.13 Bottom: The spectrum of water vapor obtained at 32 cm^{-1} resolution ("low resolution"). Top: The spectrum of water vapor obtained at 4 cm^{-1} resolution ("high resolution").

C. Determining Resolution / FTIR Trading Rules

When spectra are displayed or plotted they look like continuous functions, but in reality consist of a number of discreet data points. The computer literally plays connect the dots by drawing a line segment between each data point so the spectrum looks smooth to our eyes. The individual data points and line segments that make up a spectrum can be seen if the spectrum is greatly expanded on the computer display. Spectral resolution is a measure of the ability of an instrument to distinguish spectral features that are close together. The instrumental resolution used in measuring a spectrum determines the number of data points in the spectrum. For instance, a 4 cm^{-1} resolution spectrum contains a data point every 4 cm^{-1}, and a 32 cm^{-1} spectrum contains a data point every 32 cm-1 (ignoring the effects of zero filling). The number of data points in a spectrum can be easily calculated from the resolution and wave-

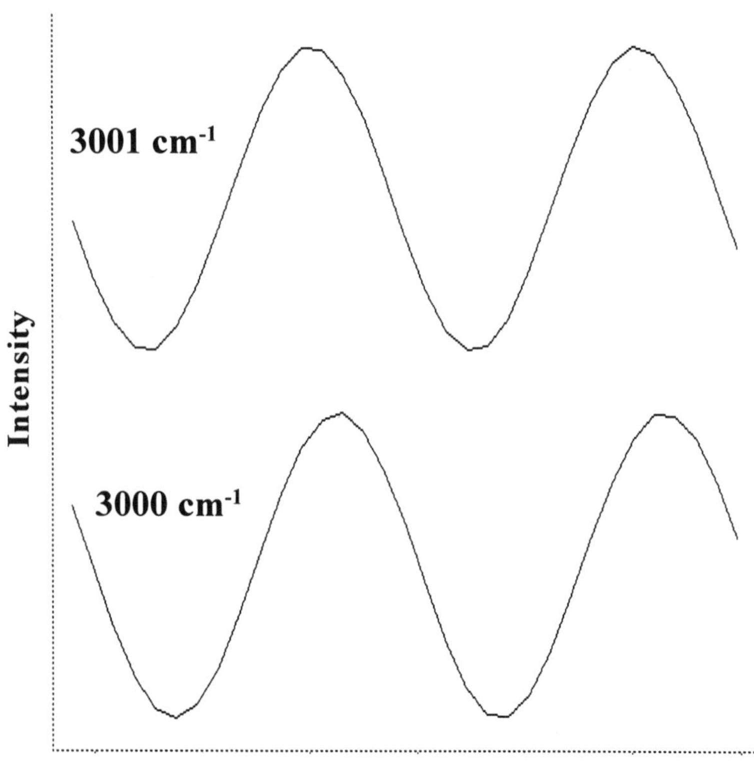

Figure 2.14 What the interferograms of light at 3000 and 3001 cm^{-1} may look like.

number range of the data. The number of data points in a 4 cm^{-1} resolution spectrum from 4000 to 400 cm^{-1} is calculated as follows:

$$4000\ cm^{-1} - 400\ cm^{-1} = 3600\ cm^{-1}\ \text{range}$$

$$3600\ cm^{-1} / (4\ cm^{-1}/\text{data point}) = 900\ \text{data points}$$

The impact of using different resolutions on the appearance of a spectrum is seen in Figure 2.13, which shows spectra of water vapor taken at 4 and 32 cm^{-1} resolution respectively. The 32 cm^{-1} spectrum is broad and featureless, while the 4 cm^{-1} spectrum shows many sharp features. The 32 cm^{-1} spectrum is said to be taken at <u>low</u> resolution because it does not distinguish spectral features well, while the 4 cm^{-1} spectrum is said to be taken at <u>high</u> resolution because it resolves many spectral features. The terms "low and "high" resolution can be confusing. Resolution is denoted by how close together in two features can be in cm^{-1} and still be distinguished. So, high resolution is denoted by a small wavenumber value, and low resolution is denoted by a large wavenumber value. The reason the low resolution spectrum looks broad and featureless in Figure 2.13 is that only features more than 32 cm^{-1} apart than can be distinguished, the data point spacing is too wide to see features any closer together. In the high resolution spectrum, features that are 4 cm^{-1} or more apart are resolved because the data point spacing is small enough to allow these features to be seen.

A more precise definition of when two bands are resolved is in order. Using the *Rayleigh criterion* [1] the valley between two adjacent peaks of equal intensity must be 20% lower than the peak tops to be considered just resolved. The reason high resolution spectra are desirable is that more spectral features can be seen in them, enhancing the information content of the spectrum.

Another way of determining the resolution of a spectrum is to look at the width of the bands. Band width is determined by measuring the *full-width at half-height* of a band, abbreviated FWHH. If a band is inherently narrow, say 0.1 cm^{-1} wide (the things that contribute to a band's inherent width will be discussed below), a 1 cm^{-1} resolution spectrum of this band should show a FWHH of 1 cm^{-1}. The FWHH in spectra of gases such as carbon monoxide (see Figure 4.19), whose bands are inherently narrow, are measured and used to determine the effective resolution of a spectrometer. For example, if a spectrum of CO is measured on an instrument, and the FWHH of the peaks is 2 cm^{-1}, the spectrum is said to be taken at 2 cm^{-1} resolution.

Resolution in an FTIR is determined optically as follows. Imagine a Michelson interferometer equipped with a broadband infrared source such as the one shown in Figure 2.1. Assume the spectrum of your sample has two absorbance bands that are close in wavenumber, say at 3000 and 3001 cm^{-1}. These bands give rise to two very similar sinusoid interferogram signals as seen in Figure 2.14. If the mirror is moved only a short distance, and only a small portion of the sinusoids are sampled, the instrument can not distinguish between the two interferograms look identical at small values of δ. Since the interferograms for light 1 cm^{-1} apart cannot be distinguished, spectral features less than 1 cm^{-1} apart cannot be distinguished. If the interferogram is measured with a large δ, the instrument will distinguish between the sinusoids and spectral features 1 cm^{-1} apart will be resolved in the resultant spectrum. This can be seen in Figure 2.14 by noting that the second crests of the two sinusoids do not overlap. Thus, as optical path difference increases the ability of the spectrometer to distinguish nearby features increases. This can be summarized in equation form as follows:

$$\text{Resolution} \propto 1/\delta \qquad (2.5)$$

Table 2.3
The Relationships between Resolution, Optical Path Difference (δ) and Mirror Displacement (Δ)

Resolution in cm^{-1}	δ in cm	Δ in cm
8	0.125	0.0625
4	0.25	0.125
2	0.5	0.25
1	1.0	0.5
0.5	2.0	1.0

where δ is the optical path difference in cm, and resolution is defined in cm^{-1}. Recall that high resolution is denoted by small numbers, which is why the relationship is reciprocal. Equation 2.5 says that to obtain a 4 cm^{-1} resolution spectrum, the optical path difference must be 1/(4 cm^{-1}) or 0.25 cm. Recall that optical path difference is twice the mirror displacement, so to achieve an optical path difference of 0.25 cm the mirror only needs to be moved 0.125 cm. Table 2.3 shows the mirror displacements and optical path differences needed to achieve specific resolutions.

The number of data points in an interferogram is determined by the maximum optical path difference used to measure the interferogram. So, long interferograms have more data points than short interferograms. This is why high resolution spectra contain more data points, and more data points per cm^{-1}, than low resolution spectra.

We need to give equation 2.5 a more thorough theoretical under pinning. Let's define the difference in wavenumber between two infrared bands as being ΔW. For the two interferograms in Figure 2.14, ΔW is 1 cm^{-1}. Now, we already know that at ZPD these two interferograms will be in phase with each other. Theory [1] tells us that these two interferogram will be totally out of phase with each other when

$$\delta = 1/2\Delta W$$

and will be totally in phase with each other when

$$\delta = 1/\Delta W$$

For the interferograms in Figure 2.14 the optical path difference would be 0.5 cm when they are totally out of phase, and 1 cm when they are totally in phase. Now, to truly resolve two spectral features, the optical path difference must be large enough so that their interferograms start in phase, go out of

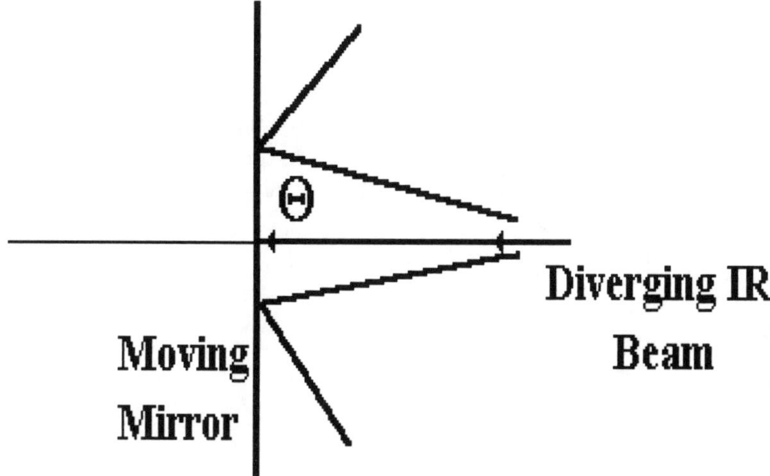

Figure 2.15 An example of the phenomenon of angular divergence. The diverging infrared beam makes an angle Θ with a line perpindicular to the surface of the moving mirror.

phase, then come back into phase with each other. This is why a minimum optical path difference of 1 cm must be used to resolve features that are 1 cm^{-1} apart.

The drawback of obtaining high resolution spectra is that they are noisier than low resolution spectra. Recall the shape of a typical interferogram, as seen in Figure 2.6. The interferogram signal is very intense at the centerburst, and decays rapidly in the wings, while the noise is constant at all points in an interferogram. For now, let's say the noise has a value of 0.05 volts. At ZPD the noise signal represents a small percentage of the total signal being measured. In the wings of the interferogram where the signal is very weak, 0.05 volts may represent a significant percentage of the total signal being measured. Short interferograms measured with small optical path differences measure less noise as a percentage of the total signal than do long interferograms measured using large optical path differences. Simply put, long interferograms measure more noise than short interferograms. This is why high resolution spectra, which are measured with large optical path differences, are noisier than low resolution spectra that are measured with small optical path differences. As a result, the relationship between SNR and resolution is as follows:

$$\text{SNR} \propto \text{Resolution} \qquad (2.6)$$

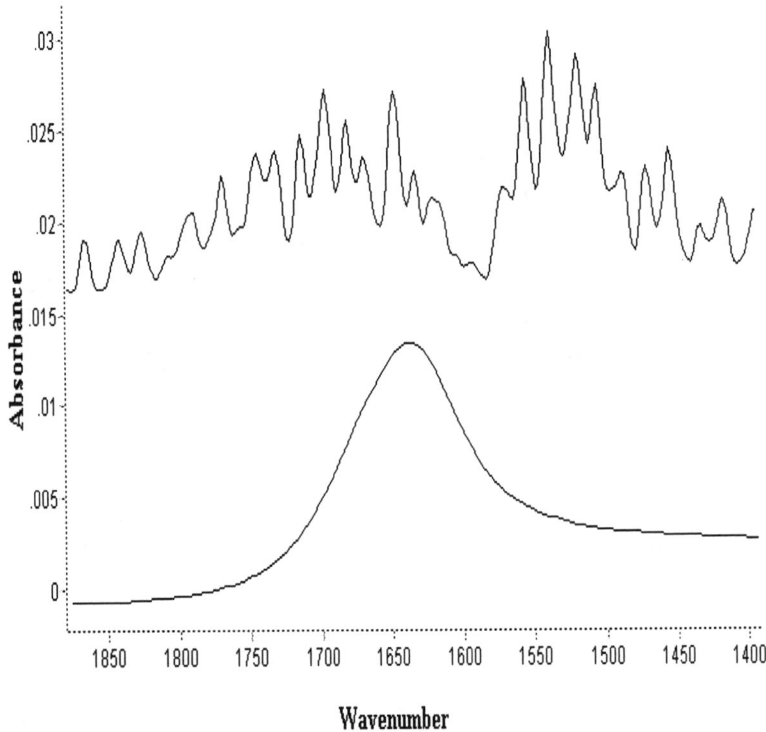

Figure 2.16 The spectra of gaseous water (bottom) and liquid water (top). Both spectra were obtained at 8 cm^{-1} resolution.

Thus, a 1 cm^{-1} resolution spectrum is lower SNR than an 8 cm^{-1} resolution spectrum. This is an important trade-off. High resolution spectra are inherently noisier, but may contain a lot of information. Low resolution spectra are not as noisy, but have a lower information content. The best resolution setting depends on your sample and what information you need to obtain from a spectrum.

There is another phenomenon that contributes to higher resolution spectra being noisier than low resolution spectra. In an ideal interferometer, the light beam is a perfectly collimated cylinder, and all light rays are parallel to each other. In reality, the optics in spectrometers are not perfect, and the infrared beam shape is not a cylinder, but a cone. The light rays in an infrared beam are not parallel, but form an angle to each other as shown in Figure 2.15. This phenomenon is known as *angular divergence.* To understand how angular divergence decreases SNR, imagine an angularly divergent beam reflecting off the moving mirror in a Michelson interferometer. Because of angular

Table 2.4
Sample Dependent Resolution Settings

Sample	Resolution
Solids, Liquids	4 to 8 cm^{-1}
Gases	4 cm^{-1} or higher

divergence, light on the outside of the beam travels a different distance than light in the center of the beam. These light beams destructively interfere with each other, reducing the amount of infrared light returning to the beamsplitter. Angular divergence increases with optical path difference because the light beam has more of a chance to spread out the further it travels. As a result, light intensity decreases as optical path difference increases, causing noise to increase with optical path difference.

An aperture can be placed in the infrared beam to counteract the effects of angular divergence. At resolutions of 4 cm^{-1} or lower an aperture is normally not needed, but at higher resolution, use of an aperture is essential. When doing high resolution spectroscopy one must choose the proper distance to move the mirror and the proper aperture size. Most high resolution FTIR systems contain computer controlled apertures that automatically adjust to the right size. A drawback of the use of apertures is that they block a portion of the infrared beam, with a concomitant loss of signal. The use of apertures is another reason high resolution spectra are noisier than low resolution spectra.

A discussion of what resolution should be used on which samples is in order. The strength and number of chemical interactions with nearest neighbor molecules varies from place to place within a chemical sample. This means there are many *chemical environments* within a sample. Molecules in each of these chemical environments absorb infrared radiation at slightly different wavenumbers, contributing to the overall width of a molecule's infrared band. So, a sample with many chemical environments has wide bands, and a sample with few chemical environments has narrow bands. An example of a sample with many chemical environments is liquid water. The strength and number of hydrogen bonds within a sample of water is variable, causing wide infrared bands as seen in the bottom of Figure 2.16. Note the band shown is over one hundred wavenumbers wide. It would not matter if this spectrum were obtained at 10 or 1 cm^{-1} resolution; the spectrum would look the same because in this case the sample is limiting the resolution, and the resolution is said to be sample limited.

Due to the large number of chemical environments in condensed phase samples, their absorbance bands are typically 10 cm^{-1} at FWHH or wider. So, a typical 4 cm^{-1} resolution FTIR spectrum will easily resolve most condensed phase sample bands. Using a higher resolution would cause the spectrum to be noisier, the measurement would take longer, and the appearance of the bands would be the same as in the low resolution spectrum. Therefore, a resolution of 4 to 8 cm^{-1} is adequate for most condensed phase samples. In certain light starved applications, such as infrared microscopy and GC-FTIR, many scans are routinely needed to obtain a good SNR. In these cases, working at low resolution allows data to be accumulated in a reasonable amount of time, so 8 cm^{-1} resolution is typically used for these techniques.

Gases exist as distinct molecules that are far apart, so the number and strength of chemical interactions between neighboring molecules is small. Thus, there are very few chemical environments in gaseous samples. Most of the molecules in a gaseous sample absorb infrared radiation at approximately the same wavenumber giving inherently narrow bands compared to condensed phase molecules. An example of a gas phase spectrum, that of water vapor, is shown in the top of Figure 2.16. The many features in this spectrum represent individual spectral lines only a few cm^{-1} wide. Note how the infrared bands of gaseous water are narrower than liquid water. When working with gaseous samples, using high resolution may show more spectral features than using low resolution. Thus, the use of high resolution on gases is justified since it may yield more spectral information than low resolution scans. Resolutions of 2 cm^{-1} or higher are routinely used to obtain gas phase spectra. If one is working with unknown gas phase mixtures, obtaining spectra at high resolution may enable the user to resolve enough lines to identify all the gases present. However, if one is working with simple mixtures of gases, a lower resolution spectrum may yield enough information for an identification or a quantitative analysis to be performed. As always, the needs of the analyst play a role in determining what resolution gas phase spectra should be obtained. A summary of typical resolution settings used for different samples is given in Table 2.4.

The ultimate resolution obtained by an instrument is not a measure of instrument quality; it is simply a measure of how far the moving mirror can be translated. If the samples you are going to normally analyze have inherently wide bands, you do not need a high resolution spectrometer. Only buy a spectrometer capable of high resolution if your samples have inherently narrow bandwidths.

The width of infrared bands is determined by a number of things, including the temperature and pressure of the sample. In the final analysis, all spectra are "sample limited" since it is the properties of the sample that determine bandwidth. However, in cases where the sample's inherent bandwidth is smaller than the resolution a spectrometer is capable of measuring, it is still appropriate to say that the resolution is spectrometer limited. The measured width of any given infrared band is determined by the instrumental resolution used and the inherent width of the band. As a general rule of thumb, the instrumental resolution should be 4 to 5 times higher than the narrowest peak in the spectrum. This insures a sufficient number of data points are measured to accurately define all the peak in a spectrum [1].

As was pointed out earlier in this chapter, when interferograms are added together the sign of random noise causes the noise to cancel itself as summarized in equation 2.4. This equation is really a special case when comparing the SNR of spectra obtained at the same resolution. It turns out that the time spent measuring interferograms is the real quantity impacting SNR. The more time spent measuring random noise, the more chances it has to cancel itself, leading to a better SNR. This is summarized as follows:

$$\mathrm{SNR} \propto (T)^{1/2} \qquad (2.7)$$

where T is the time spent measuring a spectrum. When comparing two spectra taken at the same resolution, more scans gives a better SNR than few scans because more time is spent measuring the data when many scans are coadded.

We are now in a position to understand how the multiplex advantage of FTIR (discussed in Chapter 1) really works. Imagine taking a 4 cm^{-1} resolution spectrum from 4000 to 400 cm^{-1} on a dispersive infrared spectrometer and on an FTIR. This spectrum consists of 900 data points. Assume the measurement time is 10 minutes on both instruments. The dispersive instrument measures one data point at a time, so only 1/900th of the measurement time is spent observing any given data point, which amounts to about 0.667 second. Thanks to the multiplex advantage of FTIR, all data points are measured all the time So, each data point in this spectrum is observed for a full 10 minutes (600 seconds). The SNR advantage of FTIR compared to dispersive instruments is calculated by taking the square root of the ratio of the time spent observing a data point using each instrument:

$$\text{FTIR SNR Advantage} = [600 \text{ seconds}/0.667 \text{ seconds}]^{1/2}$$

$$\text{FTIR SNR Advantage} \sim 30$$

When comparing the SNR of spectra taken at different resolutions and using different numbers of scans, the total measurement time should be used as a comparison, not the number of scans. The measurement time is easily calculated by multiplying the number of scans times the moving mirror velocity.

The trade-offs between resolution, signal-to-noise ratio, and scan time can be summarized as follows in what are known as FTIR "trading rules" [1]. When considering how to perform an FTIR analysis the parameters that need to be optimized are SNR, resolution, and analysis time. Ideally, you want a spectrum with a high SNR, obtained with a good enough resolution to resolve all the bands in the spectrum, obtained in as little time as possible. The following trading rules should always be kept in mind when setting the measurement parameters on your FTIR.

FTIR Trading Rules

1. The SNR is reduced as the square root of the amount of time spent measuring data:

$$\mathbf{SNR \propto (T)^{1/2}}$$

where SNR is signal-to-noise ratio, and T is analysis time. When comparing SNRs for spectra taken at the same resolution, the equation simplifies to:

$$\mathbf{SNR \propto (N)^{1/2}}$$

where N is the number of scans.

2. Due to electronic noise, angular divergence, and the need to use apertures, high resolution spectra are inherently noisier than low resolution spectra:

$$\mathbf{SNR \propto Resolution}$$

where SNR is signal-to-noise ratio, and resolution is measured in cm^{-1}.

3. It takes longer to obtain many scans than to obtain a few scans, so analysis time is related to the number of scans as follows:

$$\mathbf{T \propto N}$$

It takes longer to obtain a high resolution spectrum than a low resolution spectrum because it takes more time to move the mirror a longer distance. So, analysis is related to resolution as such:

$$\mathbf{T \propto 1/Resolution}$$

where N is number of scans, T is analysis time, and resolution is measured in cm^{-1}. The optimal analysis time is determined by the needs of the analyst. If a high resolution, high SNR spectrum is required, T will be large. If only basic information about a sample is needed, then lower resolution and fewer scans can be used, and T will be smaller.

In general, one should use the lowest resolution needed to obtain the required information about a sample. High resolution always involves a penalty in noise and scan time, and should only be used if the sample warrants. In conclusion, analysis time, resolution, and SNR are all related, and one must understand these relationships to optimize scan parameters to obtain high quality FTIR data.

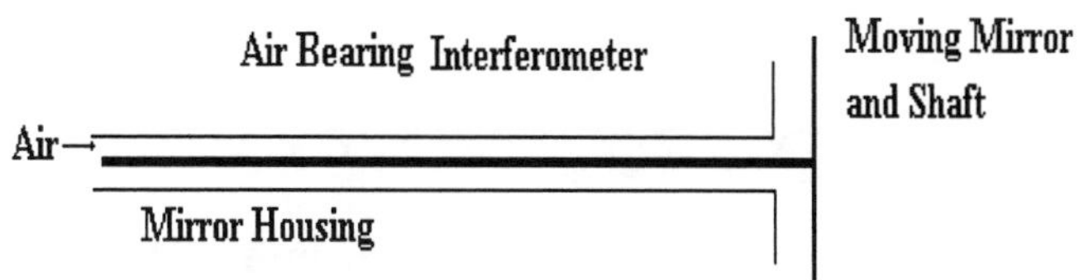

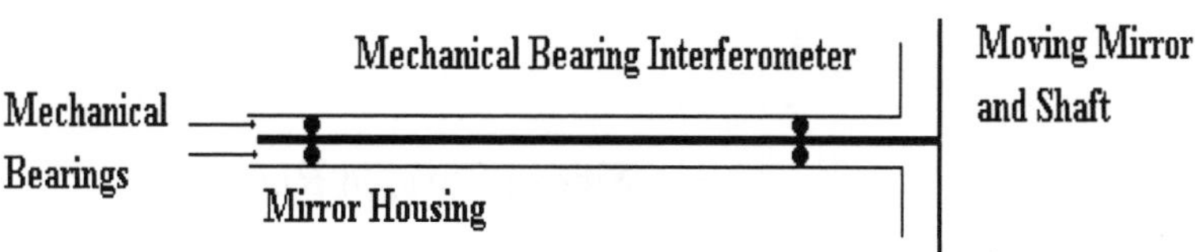

Figure 2.17 Examples of how friction between the moving mirror and its housing is handled in interferometers. Top: An air bearing. Bottom: A mechanical bearing.

D. FTIR Hardware

The wide variety of interferometers, detectors, beamsplitters, and sources available for FTIRs today can be bewildering. It is important for anyone contemplating purchase of an FTIR to know which pieces of hardware best match their applications. It is also equally important for FTIR users to know how the different parts of their instrument work together. This section discusses how commonly used interferometers, sources, beamsplitters, and detectors work, and the relative advantages and disadvantages of each.

Interferometers: Recall that the moving mirror is the only moving part in an FTIR spectrometer. It is vital that the position of the moving mirror be precisely controlled so the optical path difference can be measured accurately. The moving mirror will encounter friction with its housing. So, interferometers contain bearings to reduce or eliminate friction and to insure smooth mirror motion.

Examples of two different types of bearings used in Michelson interferometers are shown in Figure 2.17. In an *air bearing*, the shaft attached to the moving mirror is surrounded with a cushion of air upon which it floats, much like a hovercraft floats on a cushion of air. The advantage of air bearings is that they are frictionless and rarely wear out. A disadvantage of air bearing interferometers is they require a source of clean, dry air which can be expen-

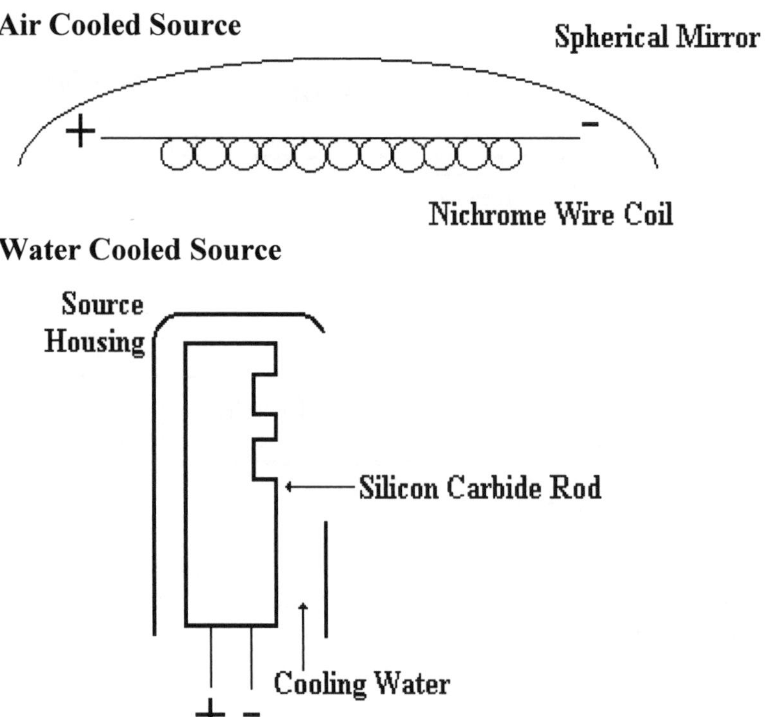

Figure 2.18 Examples of two different infrared sources used in FTIRs. Top: An air cooled source. Bottom: A water cooled source or "globar".

sive or difficult to obtain. Air bearings are also somewhat sensitive to vibrations, which can cause the shaft to bounce around and contact the housing. This leads to errors in the measurement of the interferogram. Air bearings are also expensive because they must be precisely machined.

The second way of dealing with friction in an interferometer is to use mechanical means to reduce friction, known as mechanical bearings. Ball bearings, ruby balls, and teflon pads have all been used in mechanical bearing interferometers to reduce friction. The advantage of mechanical bearing systems is they tend to be cheaper than air bearings, are less sensitive to vibrations, and require no source of dry air. The disadvantage of mechanical bearings is that they wear out. This means unpredictable amounts of down time, and the possible expense and hassle of bearing replacement.

Despite the claims of competing instrument manufacturers, every interferometer now on the market has produced thousands of quality spectra over the course of many years. Any manufacturer with an obviously deficient interferometer would have been driven out of business a long time ago. Therefore, anyone contemplating purchasing an FTIR is in the enviable position of having many quality interferometers to choose from.

Sources: The purpose of an infrared source is to provide radiant energy in the infrared region of the electromagnetic spectrum. Examples of two different types of infrared source are shown in Figure 2.18. The simplest infrared source is called an *air cooled source*. The temperature of the source is maintained by air currents in the spectrometer, hence its name. It consists of an element that may be made up of a coil of nichrome wire or a small ceramic piece. Electricity is passed through the element, and the resistance of the element to the electricity causes it to heat up and give off infrared radiation. Often times these sources are mounted with a mirror behind them to catch the radiation that would otherwise have escaped in a direction away from the sample. This helps increase the amount of infrared radiation that gets to the detector. Air cooled sources typically run at temperatures from 1100 to 1400 K. The advantages of air cooled sources is that they are inexpensive, and are convenient since they require no special cooling. However, air cooled sources may not provide enough infrared intensity for some applications.

The other major type of infrared source is the water cooled source, otherwise known as a globar. A diagram of a globar is shown in Figure 2.18. It consists of a rod of silicon carbide, and is resistively heated like the air cooled source. Water cooled sources are cooled with running water to keep them from overheating, hence their name. Globars operate at temperatures above 1400 K, so they have the advantage of giving off more infrared radiation than air cooled sources. This means the FTIR will have higher throughput and lower noise will be observed in your spectra. The disadvantage of globars is the need for cooling water, which is an added expense, and may not be available in some locations. Globars also cost more than air cooled sources.

Beamsplitters: The purpose of the beamsplitter is to split the light beam in two so some of the light reflects off the moving mirror and some of the light reflects off the fixed mirror. Potassium bromide (KBr) is almost universally used as a substrate material in FTIR beamsplitters. Although these beamsplitters are referred to as being made out of KBr, this material does not split the beam since it transmits in the infrared. Instead, a thin coating of germanium is sandwiched between two pieces of KBr, and it is this Ge coating that splits the beam. It acts similarly to a partially silvered mirror that reflects light but is still partially transparent. The KBr acts as a substrate for the beamsplitter coating,and to protect it from the environment. So, these devices are properly called Ge on KBr beamsplitters. Spectrometer companies use many proprietary coatings to enhance the optical properties of their beamsplitters. The composition of your beamsplitter coating depends on the instrument you own.

Ge on KBr beamsplitters are usable from 4000 to 400 cm^{-1}, covering the mid-infrared very well. KBr is reasonably hard, and can be manufactured to the tolerances needed for a good beamsplitter. The only drawback to KBr is that it is hygroscopic, meaning it will absorb water from the atmosphere and fog over. This is one reason why it is essential that all FTIR spectrometer benches be purged with dry nitrogen to maintain a low relative humidity, or be sealed and desiccated to keep water out altogether.

The other material typically used in mid-infrared beamsplitters is cesium iodide (CsI). This material transmits from 4000 to 200 cm^{-1}, a 200 cm^{-1} wider range than KBr. However, CsI is soft and very hygroscopic, so beamsplitters made from it must be replaced more frequently than KBr. In general, unless the spectral region from 400 to 200 cm^{-1} is very important to a user, use of CsI beamsplitters is not encouraged.

Detectors: The last mirror in an FTIR brings the infrared beam to a sharp focus on a small piece of material that known as the infrared detector element. The detector element is usually enclosed behind an infrared transparent window to protect it from the environment. The element is connected to the spectrometer's electronics, and its job is to act as a transducer; turning infrared intensity into an electrical signal. The signal is ultimately turned into a voltage, which is amplified, processed, and digitized before being Fourier transformed into a spectrum (see spectrometer electronics section below).

The most commonly used detector material in the mid-infrared is deuterated triglycine sulfate (DTGS). The DTGS detector is known as a pyroelectric bolometer, and works as follows. Changes in the amount of infrared radiation striking the detector cause the temperature of the DTGS element to change. The dielectric constant of materials such as DTGS change with temperature. The resultant change in capacitance with temperature is measured as a voltage across the detector element [2]. DTGS detectors equipped with KBr windows cover the mid-infrared range from 4000 to 400 cm^{-1}. The advantages of DTGS detectors are that they are simple, inexpensive, and robust. For the vast majority of experiments performed in the FTIR sample compartment, a DTGS detector works fine. The major drawback of DTGS detectors is that they are less sensitive than other detectors available. This means there are certain applications, such as GC-FTIR and infrared microscopy, where DTGS detectors simply cannot be used. Another problem with DTGS detectors is that they are slow, which means the moving mirror can not translate very fast while an interferogram is being measured. This means fewer scans per minute, and a longer analysis time.

The second major detector used in the mid-infrared is the mercury cadmium telluride (HgCdTe) or “MCT” detector. The MCT element consists of an alloy of these three elements, and it is a semiconductor. The detector element absorbs infrared photons, and as a result electrons are promoted from the valence band (or bonding orbital) to the conduction band (or anti-bonding orbital). Once electrons are in the conduction band they can respond to an applied voltage, giving rise to an electrical current. The electrical current is a measure of the number of electrons, and so is directly proportional to the number of infrared photons hitting the detector. Thus, the current generated by the detector element is a direct measure of infrared intensity. The energy difference between the valence band and the conduction band is called the bandgap. Photons with an energy less than the bandgap will be not be detected since they do not promote electrons to the conduction band. Photons with a an energy greater than the bandgap will promote electrons to the conduction band and will be detected. Thus, the low wavelength cutoff of these detectors is determined by their bandgaps. "Narrow band" MCT detectors are sensitive down to 700 cm^{-1}; wide band MCT detectors are sensitive down to 400 cm^{-1}. The bandgap is adjusted by changing the composition of the MCT alloy.

The major advantage of MCT detectors is their sensitivity. They are up to 10 times more sensitive than DTGS detectors. Unfortunately, there is a tradeoff between bandwidth and sensitivity with MCT detectors. The most sensitive detectors are the narrow band ones, which are useful from 4000 to 700 cm^{-1}. Wide band MCTs go down to 400 cm^{-1}, but are 5-10 times noisier than their narrow band cousins. In many applications, the wide band MCT represents only a modest improvement in sensitivity over a conventional DTGS detector. Another advantage of MCT detectors is that they are fast. As a result, one can scan many times faster than with a DTGS detector, and obtain high SNR spectra faster.

A drawback to MCT detectors is that they must be cooled, typically with liquid nitrogen. Without this cooling, heat given off by the detector element itself is detected, giving rise to a large noise signal. These detectors are equipped with dewars to hold liquid nitrogen which are easily filled, and hold enough liquid to keep the detector cool for 8 or more hours. The necessity for cooling can be a problem if liquid nitrogen is not easily available, or if one has no means of transporting and storing it. Another disadvantage of MCT detectors is their cost, which is 5 to 10 times higher than a DTGS detector. Finally, MCT detectors saturate easily. That is, if the infrared light hitting the detector element is too intense, all the available electrons will be promoted to the conduction band, and any further increase in infrared signal will give no further change in the measured electrical current. This can be a problem when doing quantitative analyses, since a linear detector response is crucial for this application. There are ways for checking for and eliminating detector saturation that will be discussed below. Certain light starved applications, such as GC-FTIR and infrared microscopy, are impossible to perform without the sensitivity of MCT detectors.

Sample Compartments: It may seem unimportant what kind of sample compartment your FTIR has, but there are two distinct types whose differences should be understood. Originally, FTIRs had sample compartments with holes in the side to let the infrared beam in and out. This maximized the amount of infrared energy that made it to the sample, and to the detector. These compartments give excellent signal-to-noise ratios, but require a purge gas to keep the KBr beamsplitter from fogging, and to prevent intense CO_2 and H_2O peaks from appearing in spectra.

More recently, manufacturers have been offering sealed and desiccated instruments. This means that infrared transparent windows, usually made of KBr, are installed over the holes in the sample compartment. The rest of the spectrometer is also sealed off, and desiccant packs that absorb water are placed therein. The advantage of a sealed and desiccated system is that no water vapor can reach the hygroscopic KBr beamsplitter. Another advantage of these systems is that they don't have to be purged with dry air or nitrogen. There may be some residual CO_2 and H_2O peaks in spectra due to the presence of these gases in the sample compartment, but the bands are usually not strong enough to be a problem. The sample compartments on these systems can be purged, but it is not a requirement. The major disadvantage of sealed and desiccated systems is that the windows partially block the infrared beam. As much as 20% of the infrared energy can be lost due to these windows, adversely affecting signal-to-noise ratios. Another disadvantage of sealed and desiccated systems is that the windows and the desiccant packs have to be replaced or regenerated regularly.

In the final analysis, the type of sample compartment to choose depends on the analyses you perform. If your FTIR is located where a dry purge gas is not available, then a sealed and desiccated spectrometer makes sense. If you are analyzing difficult samples with low signal-to-noise ratios, an unsealed system should be picked for its higher infrared throughput. If you are in neither one of these situations, then the choice of sample compartment is really one of personal preference.

E. Spectrometer Electronics

Every commercial FTIR contains a He-Ne (helium-neon) laser whose red light may be seen by holding a white sheet of paper in the infrared beam. The purpose of the laser is twofold. First, the laser acts as an internal wavenumber standard. He-Ne lasers give off light at precisely 15,798.637 cm^{-1}. All infrared wavenumbers are measured relative to it. Since we know the wavenumber of the laser line to better than three decimal places, the wavenumber reproducibility of most FTIRs is ± 0.01 cm^{-1} or better. There are some rare instrumental problems, such as laser misalignment, which can cause the accuracy of wavenumbers to be measured with an FTIR to be incorrect. One way to monitor the wavenumber accuracy of an FTIR is to occasionally obtain the spectrum of a polystyrene film (easily obtained from instrument manufacturers). Measure the position of the peak at 1601cm^{-1}. Its position should be constant. If the position of this band changes, it may mean you have problems with your instrument that my require a service call.

The second purpose of the He-Ne laser is to track the position of the moving mirror so the optical path difference can be measured properly. This is done as follows. The laser beam follows a path through the interferometer similar to that of the infrared beam. A small laser detector is mounted somewhere in the spectrometer and measures the interferogram of the laser beam. As discussed above, the interferogram of a monochromatic light source such as a laser is a sinusoid similar to the interferogram seen in Figure 2.4. The moving mirror passes through points of totally constructive interference (corresponding to maxima in the laser interferogram signal) at precise points given by equation 2.11. By using the laser's wavelength in equation 2.1, it will be found that the laser interferogram will pass through a maximum whenever the mirror moves 0.632 microns. By counting the number of intensity maxima (known as fringes) in the laser's interferogram, the precise distance the mirror has traveled can be calculated.

In an ideal spectrometer, the interferograms measured would be symmetrical about the centerburst. This means that the two wings would be identical to each other, and that only the centerburst and one wing of the interferogram need to be measured to enable a spectrum to be computed, which is known as measuring a single sided interferogram. Measuring single sided interferograms is good because it reduces the collection time by half, and the number of data points in the interferogram is reduced, increasing the speed with which the Fourier transform is calculated. In the real world, the spectrometer introduces asymmetries into the interferogram, which might necessitate measuring double sided interferograms to obtain an accurate spectrum. Asymmetric single sided interferograms are made symmetric by a process known as phase correction. Phase correction works as follows. At the beginning of a scan, a quick, low resolution, two sided interferogram (measuring both wings) is obtained. A function is calculated describing the interferogram's asymmetries. Each subsequent interferogram measured during an experiment is multiplied by this function to make it symmetric. As a result of phase correction, the advantages of collecting one sided interferograms can be enjoyed without worrying about interferogram asymmetries.

Once an interferogram is measured, it is digitized using an analog-to-digital converter (ADC). The purpose of the ADC is to turn the signal from volts (the analog signal) into a series of base 2 numbers (a digital signal), the language that a computer understands. ADCs are characterized by the number of “bits” that the digitized output contains. The largest number an ADC can measure is 2 raised to the power of the number of bits. For example, the largest number a 16 bit ADC can measure is 2^{16}, while a 17 bit ADC can measure a number as high as 2^{17}. Imagine each bit corresponds to a unit of length. Say a 16 bit ADC can measure signals corresponding to an inch in length. A 17 bit ADC would be able to measure signals corresponding to 1/2 of an inch. If your experiment produces a signal that corresponds to 1/2 of an inch, the 16 bit ADC would measure it as zero, while the 17 bit ADC would measure it as 1/2 inch. Thus, the 17 bit ADC can measure a signal that a 16 bit ADC would miss. ADCs with a high bit count are inherently better at picking up weak signals than ADCs with a low bit count. This is very important in applications where there is not much infrared signal.

The difference between the highest and lowest numbers that an ADC can measure is called its dynamic range. A large dynamic range is good because it gives one greater flexibility in the sizes of signals that can be measured, and hence more flexibility in the types of experiments that can be performed. Therefore, the main thing to remember about analog-to-digital converters is that more bits are better.

Interferograms are unusual signals in that the centerburst is very large, while the rest of the interferogram signal is very small. It is the size of the centerburst that determines how many bits of the ADC are used. Ideally, the centerburst should fill all but one or two of the bits in an ADC. For example, if the ADC can accept a signal of 10 volts maximum, a signal of about 8 volts is optimum. If the centerburst overfills the ADC, it will be truncated, and hence part of the interferogram will not be measured. This means the resultant spectrum will not be a true representation of the sample spectrum. This phenomenon is known as clipping, and is one of the major sources of noise in FTIR spectrometers. To adjust the interferogram intensity to its optimum value and avoid clipping, the entire interferogram is amplified using a gain amplifier. For lack of a better term, we will call this process gain setting. Most commercially available FTIR spectrometers make use of gain setting. In some systems, the gain is set automatically to properly fill the ADC. In other systems, the user must examine the size of the centerburst, and set the gain accordingly. Once the interferogram is digitized, it is divided by the gain setting amplification factor so the quantitative accuracy of the data is not affected. Another problem with clipping is that to reduce noise by coadding spectra, the noise must be measured in addition to the signal. If the centerburst overfills the ADC the noise signal will not be measured and coadding will not enhance SNR.

Many things are done to an interferogram before it is transformed into a spectrum. Since the intensity is weak in the wings of an interferogram, it is difficult to measure these important data points accurately. This can be overcome by amplifying the interferogram wings but not the centerburst. In a process known as gain ranging, an amplifier is turned on after a certain number of interferogram data points are collected, so the interferogram wing intensity is increased without increasing the intensity of the centerburst. The interferogram wings are sampled more accurately as a result. Most commercially available FTIR spectrometers make use of gain ranging to enhance the signal-to-noise ratios of spectra. Once the interferogram is measured, the portion that was amplified is divided by the amplification factor to preserve the quantitative accuracy of the data.

According to equation 2.3, an interferometer modulates the intensity of infrared radiation, producing a cosine wave signal at the detector. In a typical mid-infrared spectrum, the highest wavenumber of interest is 4000 cm^{-1}, which would result in an interferogram signal whose highest frequency is 8000V (where V is moving mirror velocity). Any signals in the interferogram above 8000V in frequency are not of interest, and hence are noise. These high frequency signals are eliminated from the interferogram by using an electronic low pass filter. This filter only lets signals below a certain frequency to pass. In the present example, the low pass filter would be set up to only let signals whose frequency is below 8000V to pass. The frequencies that a low pass filter must eliminate depend upon the moving mirror velocity, and the wavenumber range of interest. Therefore, the low pass filter must be programmable. In some software systems the cutoff frequency of the low pass filter is a user adjustable software parameter.

F. Measurements of Data Quality

It is important to check the performance of your FTIR on a regular basis. This provides a measure of the "health" of the instrument, and can help flag potential problems before they turn into catastrophic failures. The first thing that should be examined when using an instrument for the first time on a particular day is the interferogram. The interferogram should look similar to Figure 2.6, with a sharp centerburst dying off quickly into the interferogram wings. If the interferogram is a flat line, it probably means no signal is being measured by the detector. Check for things blocking the beam, loose cables, that the detector is chilled if it requires liquid nitrogen, or if software parameters affecting the detector are set properly.

Next, measure the size of the centerburst from its lowest to highest point. This number is called the interferogram intensity, and it is a direct measure of the total amount of infrared radiation striking the detector. Most FTIRs will show you the interferogram and measure the interferogram intensity for you. Small fluctuations in the interferogram intensity (a few percent) from scan to scan are nothing to worry about. However, rapid and large changes in interferogram intensity indicate rapid changes in the amount of infrared radiation reaching the detector. This is usually associated with the source. Perhaps the source is loose and vibrating, is about to die, or the cooling water supply is varying rapidly in flow or temperature. Long term drifts in interferogram intensity mean the interferometer has drifted out of alignment. It should be realigned per the instructions of the manufacturer.

Lastly, the X axis position of the lowest point in the interferogram should be noted. Small changes in the X axis position from scan to scan are not a problem, but large fluctuations indicate problems with the interferometer. These may be due to vibrations, moving mirror velocity errors, or problems with the He-Ne laser. Problems of this sort usually involve a call to the service department of the spectrometer manufacturer.

A background spectrum can also give information about the health of your FTIR. The shape of the spectrum and the position and intensity of the maximum in the spectrum should be reproducible from day to day. Check to see if the water and carbon dioxide peaks are extremely intense. If so, they will not ratio out well, and may mean you need to check your purge.

An important measure of how well your beamsplitter is aligned is called the beamsplitter percentage. It is calculated by taking the ratio of the maximum

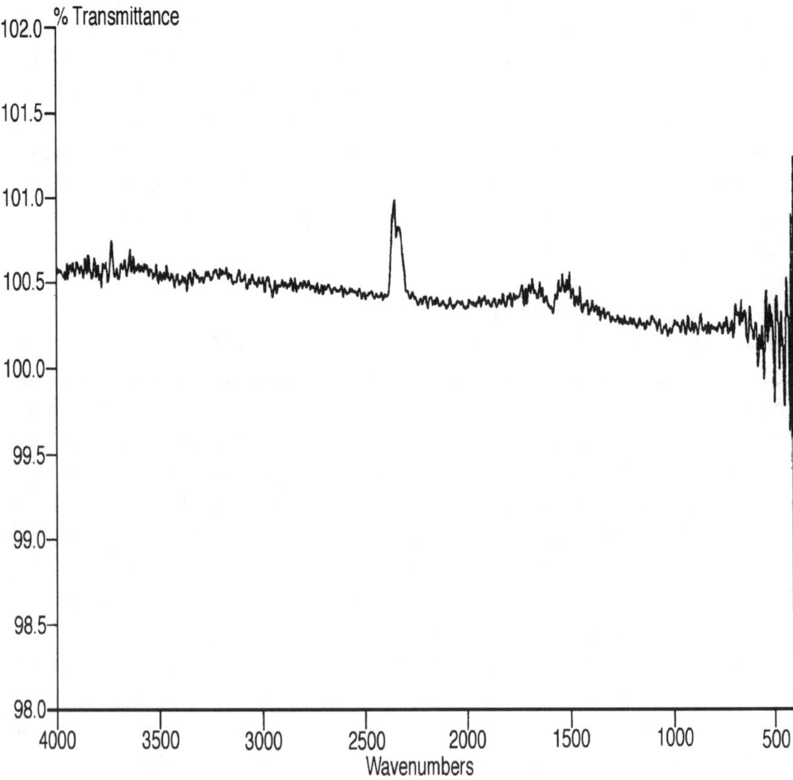

Figure 2.19 An example of a 100% line.

intensity in the background spectrum to the intensity at 4000 cm^{-1} (assuming KBr beamsplitter and DTGS detector) as follows:

$$\text{Beamsplitter \%} = I_{4000}/I_{Max}$$

where:

I_{4000} = Background spectrum intensity at 4000 cm^{-1}
I_{Max} = Maximum intensity in background spectrum

If the beamsplitter percentage is large, it means that there is a significant amount of infrared intensity at higher wavenumbers, and that the slope of the background spectrum at high wavenumber is relatively shallow. This is the ideal situation. If the beamsplitter percentage is low, it means there is very little infrared intensity at high wavenumbers, and that the slope of the background spectrum at high wavenumber is steep. This indicates that the beamsplitter may need realigning, or that it may be fogged. For the background

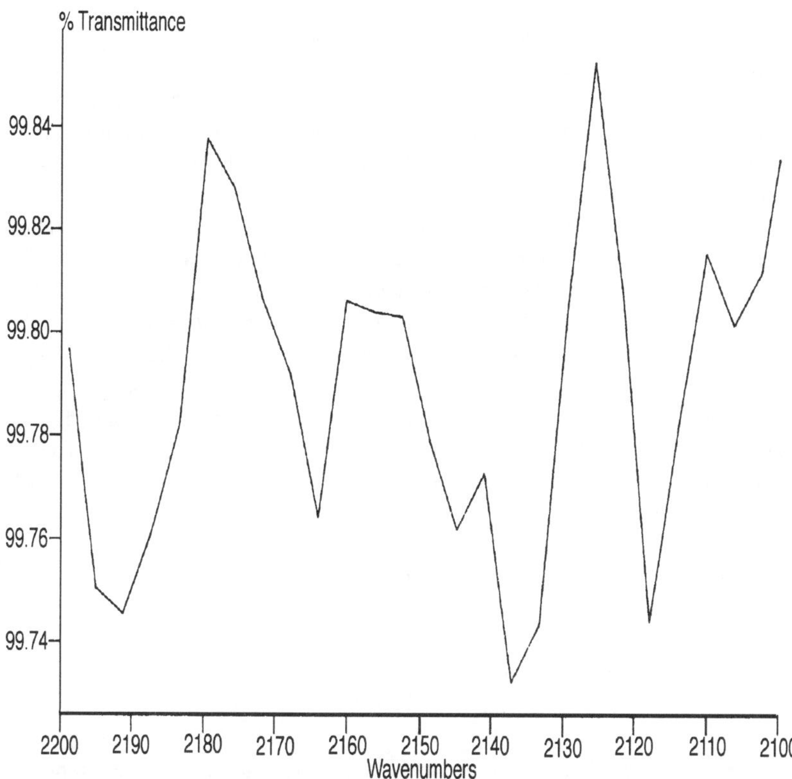

Figure 2.20 An expanded region of a 100% line.

spectrum seen in Figure 2.10 the beamsplitter percentage is 10/40 or about 25%. The exact number you obtain depends on the instrument model and manufacturer. The beamsplitter percentage should be calculated regularly, and the beamsplitter should be realigned if the percentage is too small.

The single beam spectrum should also be examined for detector saturation, which results from too much infrared radiation striking the detector. Once a detector is saturated, there is no change in detector response as more and more light strikes it. This destroys the quantitative accuracy of the spectra measured. Saturation is detected as follows. When using a KBr beamsplitter and DTGS detector, the single beam spectrum drops to zero intensity around 400 cm^{-1}. If the detector is not saturated, the single beam intensity below 400 cm^{-1} should be zero, or some very small number. If there is a rise in the single beam curve below the 400 cm^{-1} cutoff, and the intensity of this feature is greater than 5% of the maximum response in the single beam spectrum, the detector is saturating. For a narrow band MCT detector, one should look at the single beam spectrum below the 700 cm^{-1} cutoff for signs of detector saturation. Since MCT detectors are very sensitive, they saturate much more easily than DTGS detectors.

To eliminate detector saturation, the amount of infrared energy reaching the detector must be reduced. One way to do this is to place an aperture in the beam. A second way to eliminate detector saturation is by placing a neutral density filter in the beam to attenuate it. Some manufacturers provide metal mesh screens for this purpose. A crude neutral density filter can be made by cutting out a 1" square from a metal screen and placing it in the beam.

The most important measurement of FTIR performance is known as a 100% line. It is obtained by ratioing two background spectra obtained one after the other. An example of a 100% line is shown in Figure 2.19. Ideally, the 100% line should be a perfectly level line at 100% transmittance. In reality, the line in Figure 2.19 contains noise, a slight slope, and water and CO_2 bands. The slope and shape of the 100% line are very important to note. Slope can be measured by simply subtracting the %T values at 4000 and 500 cm^{-1} as follows:

$$\text{100\% Line Slope} = T_{4000} - T_{500}$$

where:

T_{4000} = Transmittance at 4000 cm^{-1}
T_{500} = Transmittance at 500 cm^{-1}

The region below 500 cm^{-1} is not used in this calculation because it tends to contain large noise spikes. For Figure 2.19, the slope is approximately

$$(100.5\%) - (100.0\%) = 0.5\%$$

Any large curvature in the 100% line indicates beamsplitter misalignment. The slope specification for a specific instrument can be obtained from its manufacturer.

A portion of the 100% line seen in Figure 2.19 is shown in Figure 2.20. The X axis is expanded between 2200 and 2100 cm^{-1}, and the Y axis has been expanded so it is full scale. What is seen in Figure 2.20 is noise, which resembles random line segments connected together. Probably the most important measure of FTIR performance is the *peak-to-peak noise*, which is the difference in the %T values for the lowest and highest points in a given wavenumber region. This region from 2200 to 2100 cm^{-1} is often used to calculate peak-to-peak noise because it is generally devoid of water vapor lines, and often corresponds to the region of highest response in the single beam spectrum. The peak-to-peak noise level in Figure 2.19 is (99.84%) - (99.74%) or 0.1% total. The noise level in the 100% line should be less than some specified value for a given resolution and number of scans. The instrument manufacturer can provide you with information on what your noise levels should be. Other ways of measuring noise, such as average or root mean square

(RMS) noise, can also be used to determine the health of an FTIR. Regardless how noise is measured, you should be consistent in the method used. Many newer spectrometers contain software programs that automatically calculate things such as the beamsplitter percentage and the noise in the 100% line. This is a convenient way to measure instrument performance. However, be careful to note how the program calculates its quantities, and remember to record the results.

Lastly and most importantly, information about interferogram intensity, beamsplitter percentage, and the 100% line should be recorded in a logbook regularly. The purpose of the logbook is to track instrument performance over time. If performance starts to slowly degrade it can flag a potential problem that can be fixed before a more catastrophic failure occurs. This can save time and money in the long run. Changes in instrument hardware and software, and any service calls should also be recorded in the logbook. Some users record all spectra obtained with an instrument in a logbook to help keep track of their work. This is encouraged.

References

1. P. Griffiths and J. de Haseth, *Fourier Transform Infrared Spectrometry*, Wiley, New York, 1986.
2. M. Diem, *Introduction to Modern Vibrational Spectroscopy*, Wiley, New York, 1993.

Bibliography

J. Robinson, *Undergraduate Instrumental Analysis*, Marcel Dekker, New York, 1995.

N. Colthup, L. Daly, & S. Wiberley, *Introduction to Infrared and Raman Spectroscopy*, Academic Press, Boston, 1990.

S. Johnston, *Fourier Transform Infrared: A Constantly Evolving Technology*, Ellis Horwood, London, 1991.

Chapter 3

Proper Use of Spectral Manipulations

A. Introduction

The manipulation of spectra is performed using the software that comes with an FTIR spectrometer, and involves altering or performing a calculation on the original spectrum received from the instrument. The purpose of manipulating a spectrum is to enhance its appearance, or extract more information from it. If spectral manipulations are not performed properly they can add artifacts to a spectrum, or completely destroy the integrity of the data. To avoid this problem, there are two rules the author has developed that you should follow whenever you perform spectral manipulations. These are known as the Laws of Spectral Manipulation.

The Laws of Spectral Manipulation

1. Always retain a copy of the unaltered spectrum, either as a data file on a computer disk, or as a plot. If a spectrum is altered or destroyed as a result of an improperly performed spectral manipulation, the original data will still be available for your use.

2. Always note, either in the spectral data file or on the spectrum's hard copy, what kind of spectral manipulations have been performed. This is vital so others who see the data will know it has been manipulated, and can take this into account when interpreting the spectrum. Also, good lab practice dictates that scientists should always report any changes made in experimental data.

Each of the following sections will discuss a specific type of spectral manipulation. The theory behind each technique will be given along with how to use each technique properly, and how to identify artifacts that have been added to the spectrum.

B. Spectral Subtraction

Spectral subtraction is performed when one desires to obtain the spectrum of a component in a mixture. Spectra to be subtracted must be in units that are linearly proportional to concentration, such as absorbance, Kubelka-Munk, or photoacoustic units. Transmission and single beam spectra should not be subtracted since their peak heights and areas are not linearly proportional to concentration. Also, spectral subtraction assumes the validity of Beer's law for the spectra involved (see Chapter 5 for a discussion of Beer's law).

When performing a subtraction, the spectrum of the mixture is called the *sample*, and the spectrum of a component to be subtracted from the sample spectrum is called the *reference* (sometimes called the subtrahend). The principle behind spectral subtraction is simple: the absorbance values of the reference are subtracted point-by-point from the absorbance values of the sample. For example, if the absorbance of the sample at 3400 cm^{-1} is 0.7, and the absorbance of the reference at the same wavenumber is 0.4, the subtraction result at 3400 cm^{-1} would be 0.3 absorbance units.

Performing subtractions in such a direct manner ignores the fact that the concentration of the reference material may have been different in the sample and reference spectra. For example, the concentration of water in its pure form is higher than in a soap and water solution. This is because the water is actually diluted when the soap is dissolved in it. To perform the subtraction without taking into account the concentration differences of water in the two spectra would mean that there would be water bands in the result. To get around this problem the reference spectrum is multiplied times a *subtraction factor*. The general equation used to perform subtractions is as follows:

$$\text{(Sample)} - \text{(Subtraction Factor)} * \text{(Reference)} = \text{Result} \qquad (3.1)$$

The absorbance values of the reference spectrum are multiplied by the subtraction factor (also called the scale factor), then subtracted point by point from the absorbance values of the sample spectrum. Best results are obtained when the absorbances of the sample and the reference are about the same, and if a subtraction factor close to one is used. The result of a subtraction will hopefully be devoid of any spectral features due to the reference.

Before performing a subtraction, visually compare the sample and reference spectra. Bands common to both spectra are due to the reference material, while bands that appear only in the sample are of interest and should be preserved in the subtraction result if at all possible. The subtraction factor is the only user adjustable parameter in the subtraction process. It is not always obvious what value of the subtraction factor will give the best results. Two steps to follow to help in this process are

1. Choose a spectral feature common to both the sample and reference spectra that is less than 0.8 absorbance units. Bands more intense than this may not

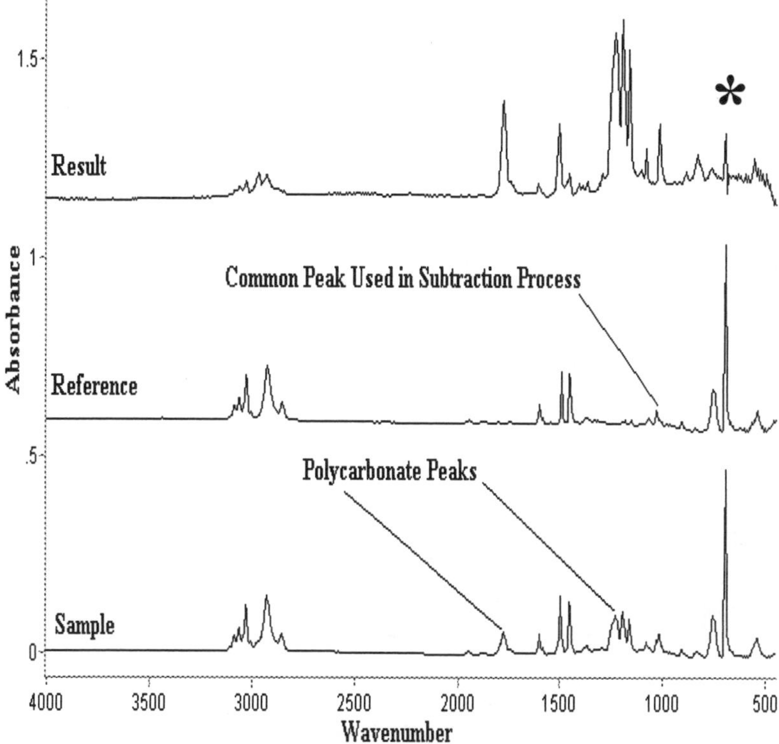

Figure 3.1 Bottom: A sample spectrum, which is a mixture of polystyrene and a polycarbonate. Middle: The reference spectrum of pure polystyrene. Top: The result of subtracting the middle spectrum from the bottom spectrum using a subtraction factor of 1.7. The peak marked with an asterisk in the result is an unsubtracted reference band with a derivative shape.

follow Beer's law, while bands at this intensity or lower usually do follow Beer's law. If a band follows Beer's law it is more likely to subtract out properly.

2. Adjust the subtraction factor interactively until the feature chosen in step 1 appears flat and becomes part of the baseline.

If there are several absorbance bands that meet the criteria stated in step 1, you can try using each band in turn to adjust the subtraction factor. It may turn out that using one band may give better results than another. Also, a subtraction factor set by using several bands at a time may give better results

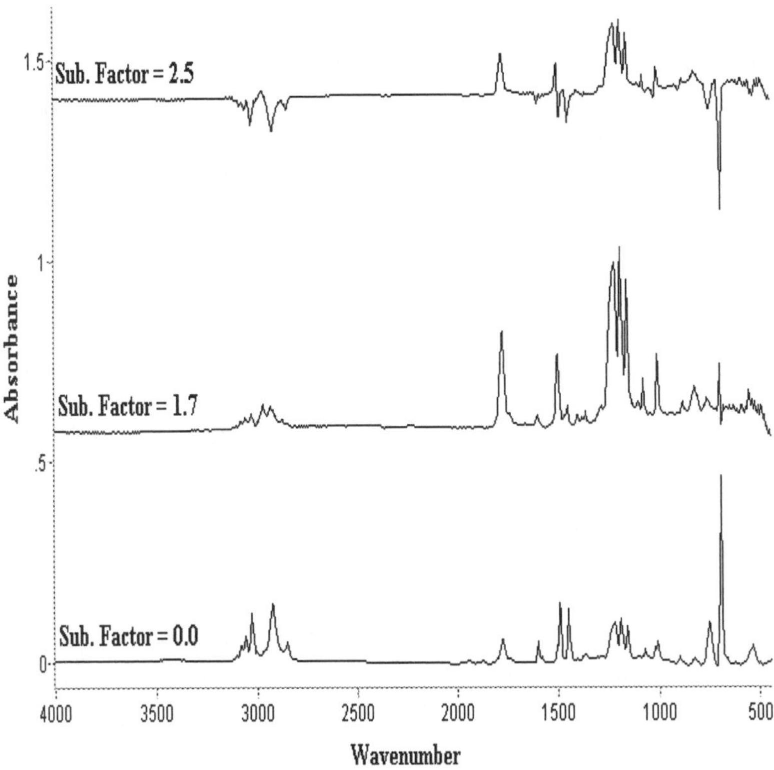

Figure 3.2 The results of using different subtraction factors for the sample and reference spectra seen in Figure 3.1. A subtraction factor of 1.7 gives good results, but subtraction factors of 0.0 and 2.5 do not.

than using just one band.

The process of subtraction is illustrated in Figure 3.1. The bottom spectrum in the figure is of a polystyrene/polycarbonate mixture, the middle spectrum is of pure polystyrene, and the top spectrum is of the subtraction result. There are bands in the sample spectrum around 1700 and 1200 cm^{-1} that are from the polycarbonate, but there is not enough information present to determine the structure of the polycarbonate. The feature in the polystyrene spectrum at 1029 cm^{-1} was chosen to be nullified because it is common to both spectra and is less than 0.8 absorbance units high. The subtraction factor was adjusted to a value of 1.7, at which point the peak at 1029 cm^{-1} was no longer visible. The result shows many bands between 2000 and 800 cm^{-1} that are definitely due to the polycarbonate, and would be very useful in determining its molecular structure. The peak marked with an asterisk in the top of Figure 3.1 is an unsubtracted polystyrene band.

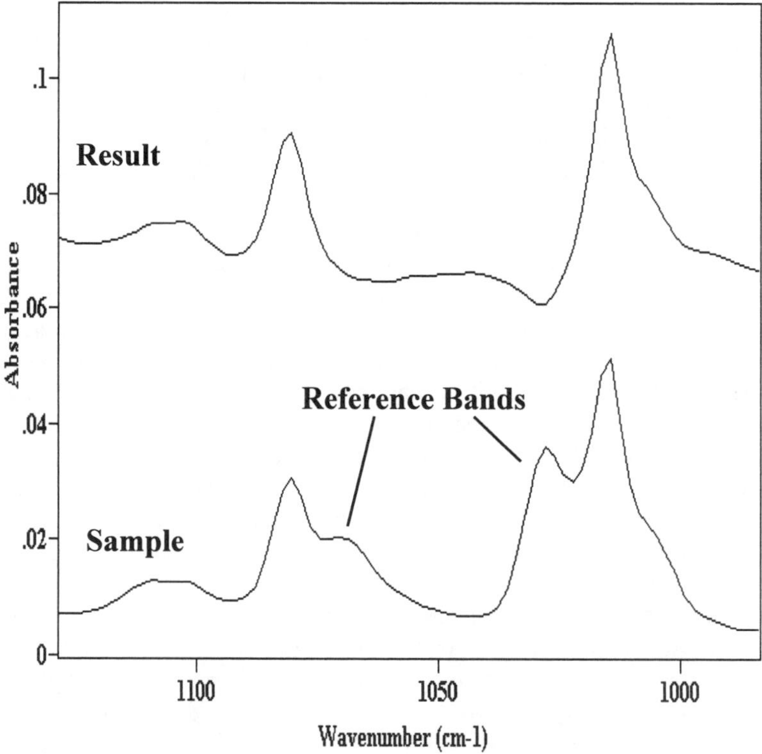

Figure 3.3 Bottom: An expansion of the sample spectrum from Figure 3.1. Note the reference bands indicated. Top: A portion of the subtraction result showing polycarbonate peaks at 1080 and 1014 cm^{-1}. These peaks were obscured by reference bands in the sample spectrum, but show up clearly after the subtraction.

Figure 3.2 shows a graphic example of the subtraction process. The spectra in this figure are subtraction results obtained with the sample and reference spectra seen in Figure 3.1, but using different subtraction factors. Note that using too small of a subtraction factor results in the reference bands still being very noticeable. Using a subtraction factor that is too large causes many downward pointing bands. Using the right subtraction factor gives a spectrum mostly free of reference bands, yet clearly shows many sample bands.

Figure 3.3 is an expansion of the subtraction result, and shows there are polycarbonate bands at 1080 and 1014 cm^{-1}. These were partially obscured by polystyrene peaks in the sample spectrum, but are plainly visible after the subtraction. This shows the ability of subtraction to find peaks of interest even if they are obscured by peaks from the reference material.

Ideally, a subtraction result should contain no bands from the reference spectrum, be free of artifacts, and have a flat baseline. In reality, there are two types of spectral artifacts that may appear in subtracted spectra. The first can be seen by examining the peak marked with an asterisk in the top spectrum of Figure 3.1. This peak points above and below the baseline, and is similar in appearance to the derivative of a spectrum. Peaks with this shape are called *derivative shaped peaks*. These artifacts are caused by wavenumber shifts in the spectrum of the reference material versus its spectrum when in a mixture. For example, pure water absorbs strongly around 1630 cm^{-1}. The presence of solutes in water may give rise to chemical interactions that can cause the 1630 cm^{-1} band maximum to shift to 1626 cm^{-1}. These chemical interactions can also cause the water bands in the sample to broaden compared to the bands in the reference spectrum. Either way, when the water reference spectrum is subtracted from a sample spectrum containing water, the bands will not overlap properly and therefore will not subtract cleanly. Part of the band will be subtracted, while part of it will remain. The result is a derivative shaped peak. Since nothing can be done to avoid the chemical interactions that cause band positions to move, nothing can be done about derivative shaped peaks. You simply need to recognize and ignore them.

The second artifact that can appear in subtracted spectra are peaks from the reference material that were not subtracted out completely. Several of these features are seen in the top spectrum of Figure 3.2. Peaks that are not subtracted properly may point up or down. The peak heights of these bands do not follow Beer's law, which is why they appear in the result. No amount of manipulation of the subtraction factor will eliminate the presence of these peaks. Reducing the concentration or pathlength of the sample may bring all peak absorbances into a range where they may follow Beer's law, and hence will subtract out. Otherwise, examine the reference spectrum to determine where intense absorbance bands occur, and ignore these bands. Use regions of the result that have no strong reference bands in them for spectral interpretation.

Very few subtractions are perfect. However, by following the process outlined above, and by being aware of the existence of artifacts that can be generated as a result of performing a subtraction, this form of spectral manipulation should yield useful information for you.

C. Baseline Correction

Baseline correction is used to correct spectra that have sloping or curved baselines so the result is a spectrum with a flat baseline. There are many reasons why a spectrum's baseline may not be flat, including sample scattering, inappropriate choice of background, and instrument drift.

Baseline corrections are performed by first generating a function (either a straight line or a curve) that parallels the shape of the baseline in the sample spectrum. The function can be generated in a number of different ways, depending on the software package used. The function is then subtracted from

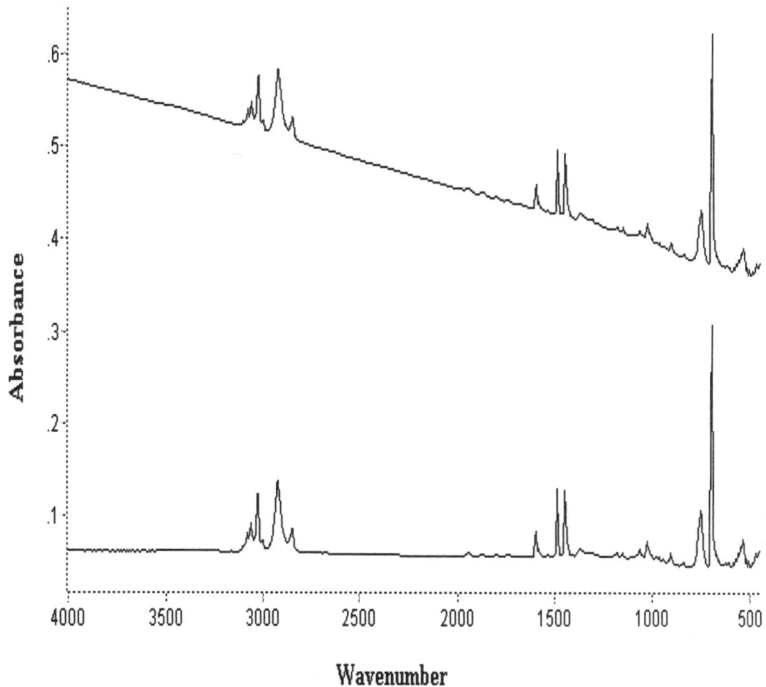

Figure 3.4 Top: An example of a spectrum with baseline slope in need of baseline correction. Bottom: The same spectrum after baseline correction. A function consisting of line segments parallel to the slope in the baseline was subtracted to effect the correction.

the sample spectrum, yielding a result that should be devoid of the curved or sloped baseline.

An example of successful baseline correction is shown in Figure 3.4. The top spectrum is of a polystyrene film with a sloping baseline. A baseline correction program was used to generate a function parallel to the baseline consisting of 12 line segments, which was subsequently subtracted from the sample spectrum. The result of the baseline correction is shown at the bottom of Figure 3.4. The baseline is flat, and there are no apparent spectral artifacts introduced by the baseline correction process.

It is vital that the function subtracted from the sample spectrum be parallel to the baseline; or artifacts will be introduced into the corrected spectrum. An example of what can go wrong with a baseline correction is shown in Figure 3.5. The bottom spectrum in the figure is a mixture of water vapor and methane. The spectrum contains a curved spectral artifact at 600 cm^{-1}. A straight line was drawn between the end points of the curved artifact and subtracted from the spectrum. A straight line does not parallel a curve very well, so the

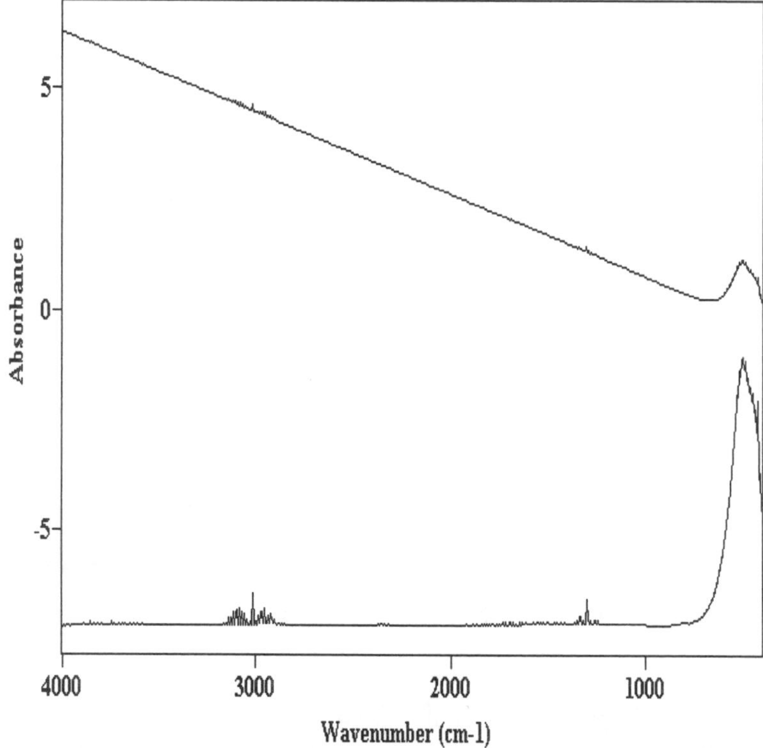

Figure 3.5 Bottom: A spectrum with a large artifact at low wavenumber in need of baseline correction. Top: An example of an improper baseline correction performed on the bottom spectrum.

result shown in the top of Figure 3.5 looks terrible. The spectral artifact now looks real, while the rest of the spectrum has been skewed and has a slope. The proper way to correct this spectrum would be to use many line segments to approximate the shape of the curved artifact, or to use a program that allows you to draw and subtract a curve from the spectrum.

The vast majority of baseline correction programs work by allowing the user to choose the data points between which to draw line segments. The choice of line segment end points is very critical. For absorbance spectra, local minima should be used. For transmittance spectra, local maxima should be used. Also, you should use the minimum number of line segments that produces a function that parallels the baseline well. The fewer segments, the less the impact baseline correction has on the appearance of the peaks in the corrected spectrum. In general, baseline correction is a useful and easily performed type of spectral manipulation, as long as the problems discussed above are avoided.

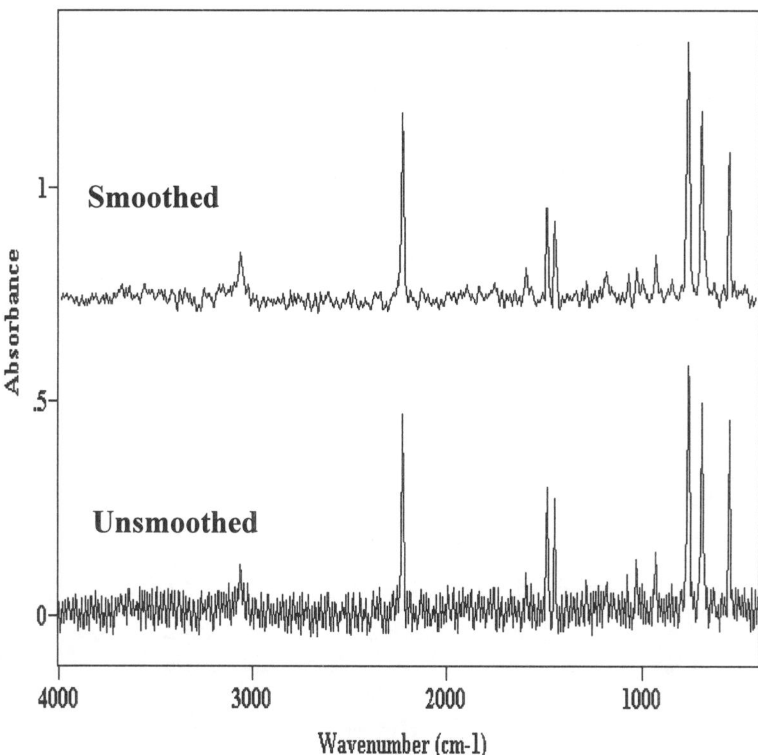

Figure 3.6 Bottom: An example of a noisy spectrum. Top: The same spectrum after a 9 point Savitsky-Golay smooth. Note the reduction in noise, and the appearance of absorabance bands previously unobservable.

D. Smoothing

Smoothing is used on noisy spectra to reduce the noise level, so features that may have been hidden under the noise can be seen more readily. Thus, smoothing enhances the information content of a spectrum. Smoothing also has a cosmetic effect, improving the overall appearance of bad looking data. The effects of smoothing on noisy data can be seen in Figure 3.6. The bottom spectrum in the figure contains a large amount of noise. The top spectrum is the result of performing a 9 point Savitsky-Golay smooth. Note that the overall noise level has gone down, and that absorbance peaks that were previously difficult to see are now easily seen.

The simplest method of performing smoothing, called a boxcar smooth, works as follows. Imagine taking several data points in a spectrum, and drawing a box around them. This box is known as the *smoothing window.* Assume the smoothing window contains 9 data points. Now, take the Y axis values (ei-

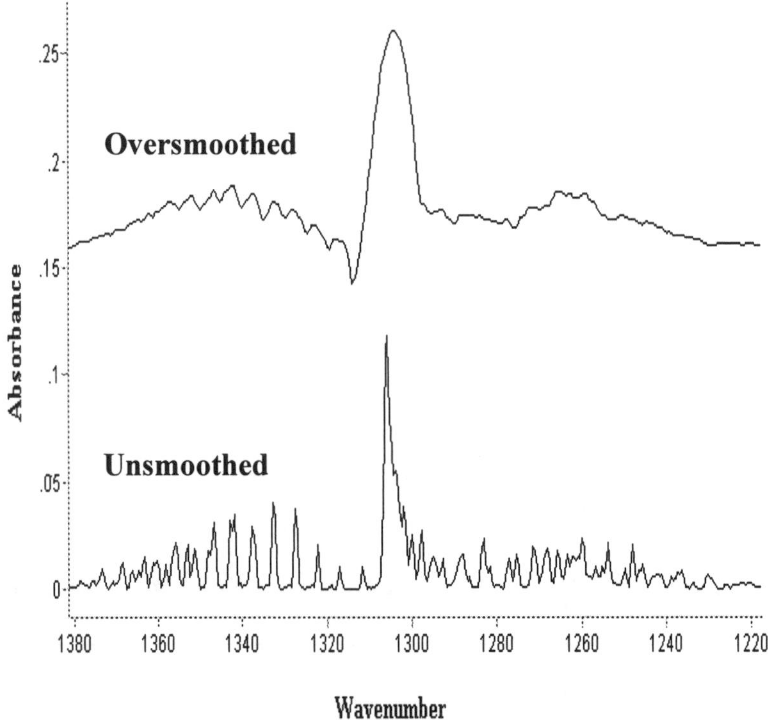

Figure 3.7 An illustration of the effects of oversmoothing. Bottom: The spectrum of methane obtained at 1 cm^{-1} resolution. Top: Same spectrum after a 35 point Savitsky-Golay smooth.

ther absorbance or transmittance) of these 9 data points, add them together, then divide by 9 to calculate the average of the values. Assign the average Y axis value to the middle of the 9 X axis data points. Next, move the box over one data point, calculate the average for this set of 9 data points, assign a new average to the middle X axis data point and so on. Eventually the smoothing window is passed over the entire spectrum to obtain a running average of the Y axis values. The result is the average for each set of data points plotted against wavenumber. The total number of data points in the spectrum is affected slightly depending on the smoothing algorithm used. The noise in the smoothed spectrum is reduced because fluctuations in the sign of random noise cancel out when the Y axis values are averaged (for a more detailed discussion of how random noise cancels itself, see Chapter 2). The amount of smoothing is proportional to the number of points included in the smoothing window. The more points in the window, the greater the extent of the smoothing.

There are different ways of calculating the average of the data points in the smoothing window called *smoothing algorithms*. There may be several different smoothing algorithms available, depending on your software package. For instance, averages that apply more or less weight to midpoints and end points in the smoothing window can be used. Some of the more common smoothing algorithms are boxcar, triangle, fast Fourier transform, and Savitsky-Golay. Terms such as “7 point boxcar” and “9 point Savitsky-Golay” are proper ways of denoting how a spectrum has been smoothed. The Savitsky-Golay smooth is probably the best known, and works by fitting a polynomial function to the set of data points in the smoothing window [1]. The higher the degree of the polynomial used, the lower the amount of smoothing achieved.

The drawback of smoothing is that it causes apparent spectral resolution to be reduced. This is a natural outcome of the averaging process. If smoothing is not performed properly, and *oversmoothing* occurs, narrow spectral features will be deresolved and grow together to become one feature, or a single narrow feature might be destroyed altogether. An example of oversmoothing is shown in Figure 3.7. The bottom spectrum is of methane obtained at 1 cm^{-1} resolution. The top spectrum is the result after a 35 point Savitsky-Golay smooth (with a polynomial value of 2). Note that the many sharp features in the original spectrum have grown together to form a few wide bands.

To prevent oversmoothing from affecting your data, smooth in small increments, say 5 to 9 points at a time. Keep a close eye on the narrowest feature that was in the unsmoothed spectrum. If this feature starts to disappear, or grow into any other features around it, you have oversmoothed. One should be careful whenever using this spectral manipulation technique.

E. Spectral Derivatives

As you may recall from calculus, the slope of any mathematical function can be determined by calculating its *derivative*. Since an infrared spectrum is a mathematical function, its derivative can be calculated. The derivative of a spectrum can be taken a number of times, producing derivatives of different orders. For instance, the first derivative of a spectrum is called a first order derivative, the derivative of the first derivative is called a second order derivative and so on. The simplest way to calculate a spectral derivative is the "point difference" method, where the difference in Y values between successive data points is calculated, then plotted versus wavenumber. There are other ways of calculating derivatives, depending on the software package you use.

An example of a first derivative is shown in Figure 3.8. There is one upward pointing and one downward pointing feature in the derivative for each absorbance peak in the original spectrum. The wavenumber at zero absorbance in a first derivative corresponds to the wavenumber at the top of the absorbance band in the original spectrum. As a result, derivatives help pinpoint the peak position of a band.

Second derivatives are calculated by taking the derivative of a first deriva-

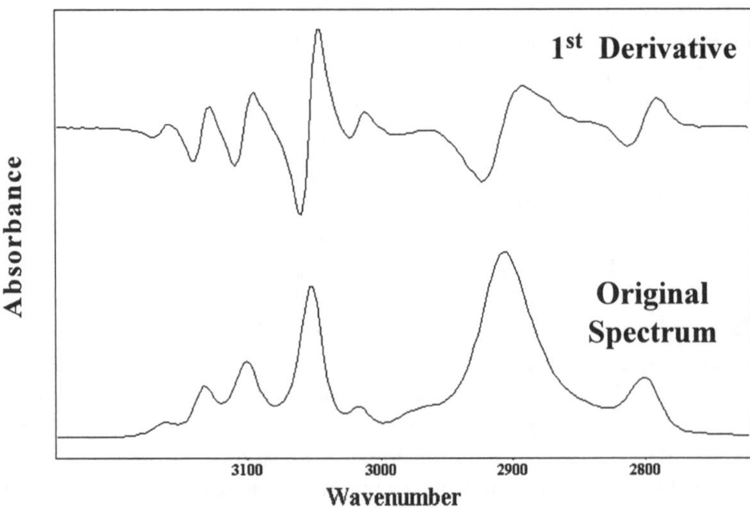

Figure 3.8 Bottom: The spectrum of polystyrene around 3000 cm^{-1}. Top: The first derivative of this spectrum. Note there is one upward and one downward pointing feature for each band in the original spectrum.

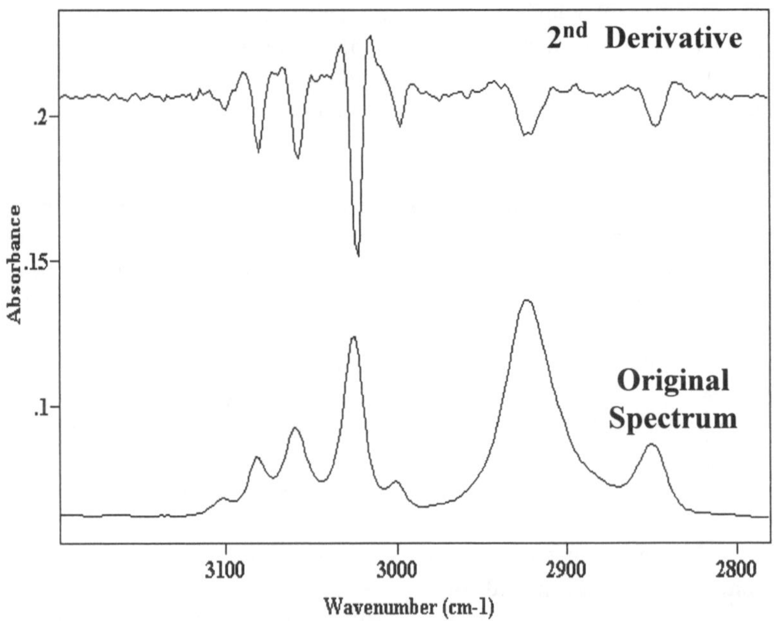

Figure 3.9 Bottom: The spectrum of polystyrene around 3000 cm^{-1}. Top: The second derivative of this spectrum. Note that there is one downward and two upward pointing features in the derivative for each absorbance band in the spectrum.

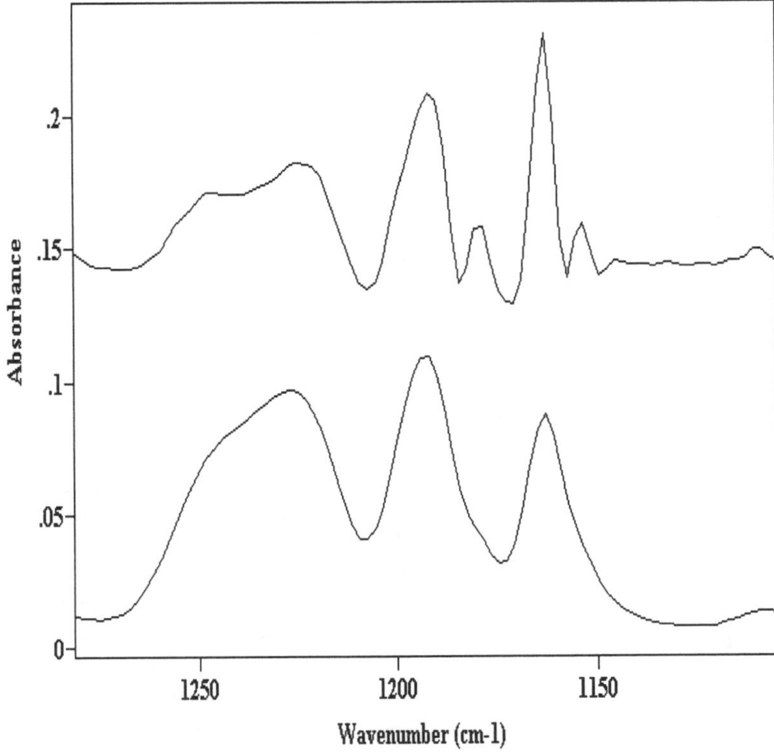

Figure 3.10 Bottom: A portion of a polystyrene/polycarbonate spectrum consisting of several overlapped bands. Top: After deconvolution. Note the sharper bands and the appearance of new features in the deconvolved spectrum.

tive. An example a second derivative is shown in Figure 3.9. The second derivative contains three features corresponding to each absorbance peak in the original spectrum, two pointing upward and one pointing down. The bottom of a downward pointing feature in a second derivative corresponds exactly to the wavenumber of maximum absorbance of a band in the original spectrum. This is why second derivative spectra are used in library searching and peak picking. If there is a region of a spectrum where several bands have overlapped to form one broad band, the number of downward pointing peaks in the second derivative gives a good estimate of the number of overlapped bands in the region. As a result, the second derivative of a spectrum should be examined prior to deconvolution or curve fitting (see below).

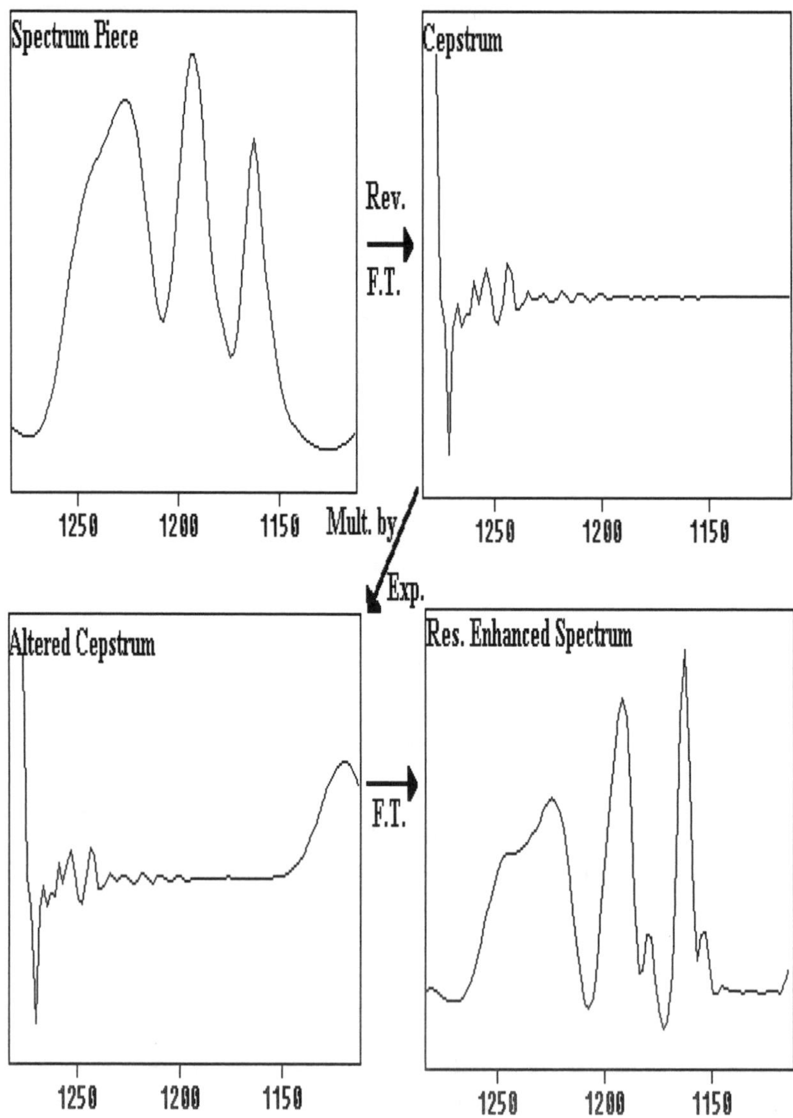

Figure 3.11 How the deconvolution process works. A piece of a spectrum (upper left) is reverse Fourier transformed to yield a cepstrum (upper right). The cepstrum is multiplied by an exponential function to increase the intensity in the cepstrum's wings to yield an altered cepstrum (lower left). The altered cepstrum is Fourier transformed to give the resolution enhanced spectrum (lower right).

F. Deconvolution

The purpose of deconvolution is to mathematically enhance the resolution of a spectrum [2]. It is most commonly performed on a section of a spectrum where several narrower bands overlap to give a broad band. Deconvolution can help determine the number and positions of the overlapped bands. The process of deconvolution retains peak position information, but peak areas and peak shapes are altered in the process. Therefore, it is important to not perform quantitative analysis using deconvolved spectra.

An example of a successful deconvolution is shown in Figure 3.10. The bottom is a section of the spectrum of a polycarbonate/polystyrene mixture. These bands are broad, and the band at 1220 cm^{-1} contains a shoulder, suggesting the presence of at least two bands underneath it. The polycarbonate spectrum was successfully deconvolved as seen in the top of Figure 3.10. There are 6 bands in the deconvolved spectrum versus three in the original data. Using guidelines discussed below, each one of these features was determined to be real. The power of deconvolution is that these new features may contain important information about a sample's composition.

The fact that resolution can be enhanced mathematically does not mean we have violated the rule that the resolution of a spectrum is determined by the maximum optical path difference achieved during a scan. Deconvolution is used on bands whose widths are inherently broader than the instrumental resolution used. The fundamental limit on the amount of resolution enhancement that can be achieved with deconvolution is the instrumental resolution used to obtain the data. There are many other limitations on the amount of resolution enhancement to use, as discussed below.

The entire deconvolution process is illustrated in Figure 3.11. The first step in the process is to choose a section of the spectrum containing bands that need to be resolved. Such a spectrum is seen in the upper left of Figure 3.11. A reverse Fourier transform is performed on this part of the spectrum to produce a result called a *cepstrum* (spectrum spelled sideways?) as seen in the upper right of Figure 3.11. A cepstrum is very similar to an interferogram, except that it corresponds to a piece of a spectrum rather than to a whole spectrum. Cepstrums have some unique properties as illustrated in Figure 3.12. Narrow infrared bands give cepstrums with significant amplitude in the wings, while wide infrared bands give cepstrums with low amplitude in the wings.

The idea behind deconvolution is to make the cepstrum of a wide band look like the cepstrum of a narrow band. This is done by increasing the wing intensity of the cepstrum obtained from a wide band by multiplying it by an exponential function $e^{\alpha X}$, where X is optical retardation and α is related to the amount of resolution enhancement. The Y value of exponential functions increases rapidly as X increases, so after multiplication the wings of the cepstrum are greatly enhanced as seen in Figure 3.11 to produce an altered cepstrum. The altered cepstrum is Fourier transformed to obtain the resolution enhanced spectrum as seen in the bottom right of Figure 3.11. The altered

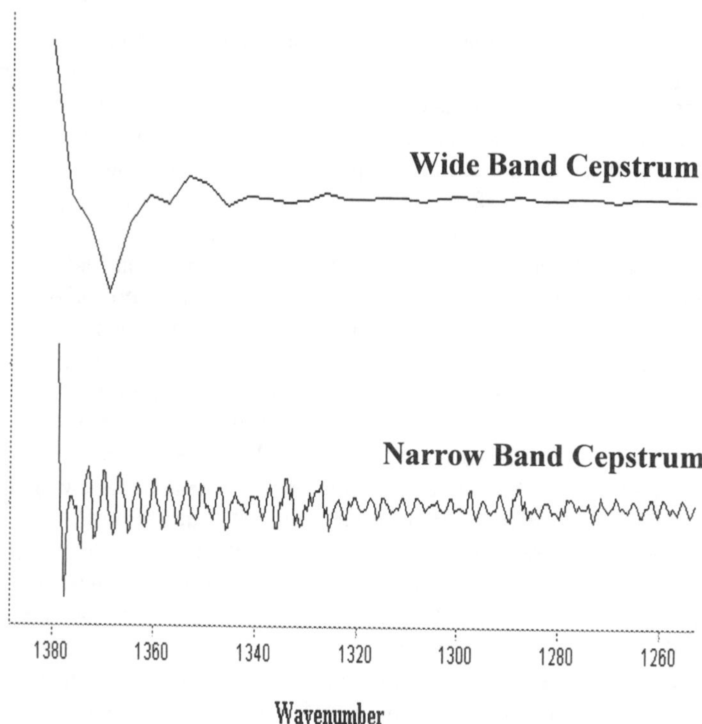

Figure 3.12 Bottom: The cepstrum of a narrow infrared band. Note how the intensity does not drop off quickly. Top: The cepstrum of a broad infrared band. Note how the intensity drops off quickly.

spectrum ideally has more and/or narrower bands than the unaltered spectrum.

The most important user adjustable parameter in deconvolution is the amount of resolution enhancement. Unfortunately, different software packages have different names and values for the resolution enhancement factor. The simplest way to calculate a resolution enhancement factor on your own is to choose a peak in the region to be deconvolved, and measure its FWHH (see chapter 2). After deconvolution, measure the width of the same peak again. The ratio of the widths before and after deconvolution is the resolution enhancement factor.

It is not obvious from looking at a spectrum how much resolution enhancement is appropriate. One can enhance the resolution of a spectrum until hundreds or thousands of new features appear, resulting in *overdeconvolution.* An example of overdeconvolution is shown in Figure 3.13. The bottom spectrum in the figure is the result of a proper deconvolution. The top spectrum is

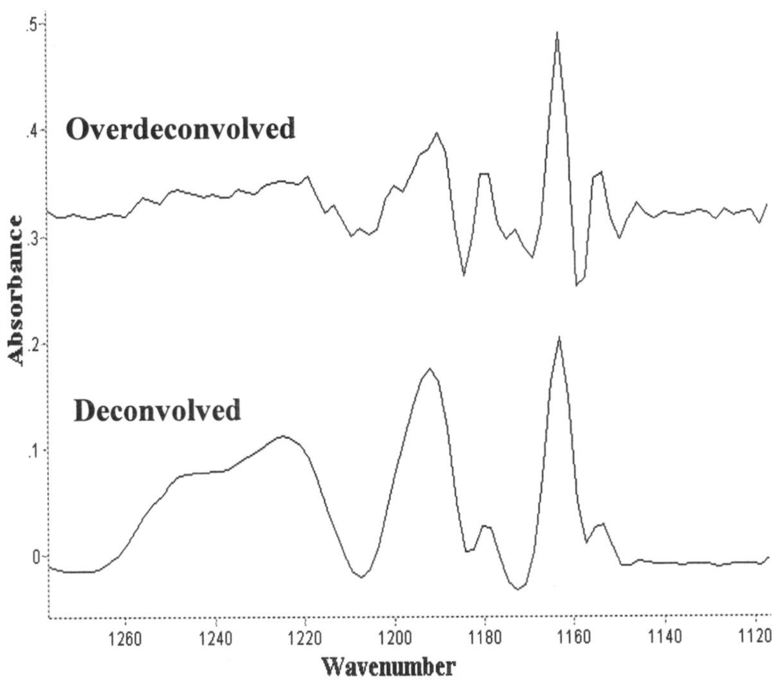

Figure 3.13 Bottom: A properly deconvolved spectrum. Top: An overdeconvolved spectrum exhibiting noise and baseline undulations.

Table 3.1
Maximum Resolution Enhancement That Should Be Used in Deconvolution Given a Specific Signal-to-Noise Ratio

SNR	Max. Res. Enhancement
100:1	2
1000:1	3
10,000:1	4

very overdeconvolved, showing noise spikes and baseline undulations. In fact, the overdeconvolved spectrum bears little or no resemblance to the properly deconvolved spectrum.

To avoid the problems of deconvolution, there are several guidelines (in addition to the instrumental resolution guideline mentioned above) that can help determine the appropriate amount of resolution enhancement to use. Since deconvolution enhances the noise in a spectrum, one should only deconvolve

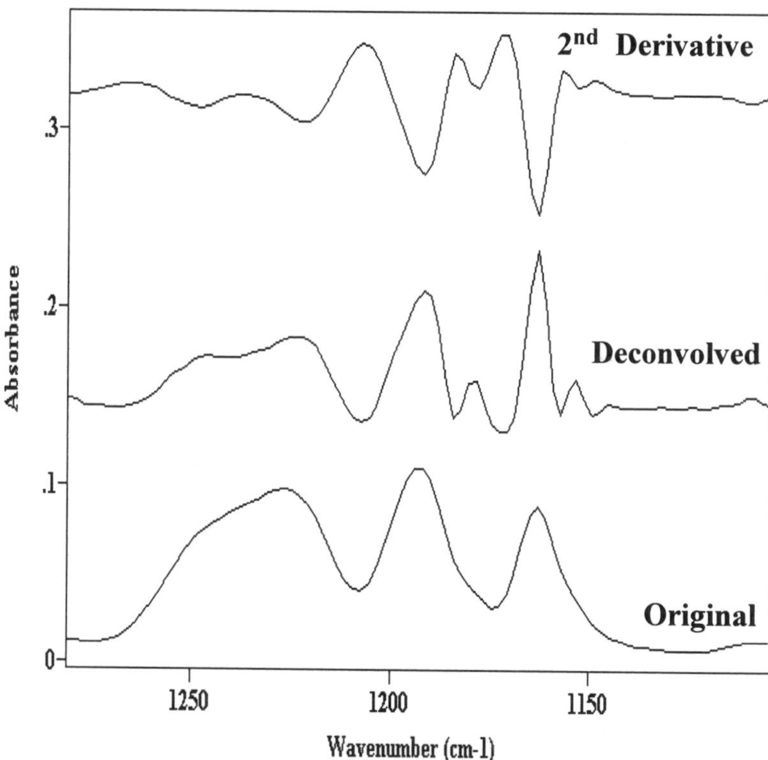

Figure 3.14 Bottom: A portion of a polycarbonate/polystyrene spectrum. Middle: The same spectrum deconvolved. Top: The second derivative of the spectrum. Note that for each downward pointing feature in the second derivative there is a corresponding absorbance band in the deconvolved spectrum.

high signal-to-noise ratio (SNR) spectra. There is an exponential relationship between the amount of resolution enhancement and SNR as shown in Table 3.1. The table gives the maximum resolution enhancement that should be used on a spectrum with a given signal-to-noise ratio.

The best guide to how much resolution enhancement to use is to deconvolve until the number of real features you expect to see in a given wavenumber range appears. There are several ways of determining how many real features have grown together to make up an overlapped band. A simple inspection of the spectrum may show shoulders and inflection points that may indicate the approximate number of overlapped bands. However, the best method for determining the number of bands in a region is to take the second derivative of the region. As discussed earlier in this chapter, second deriva-

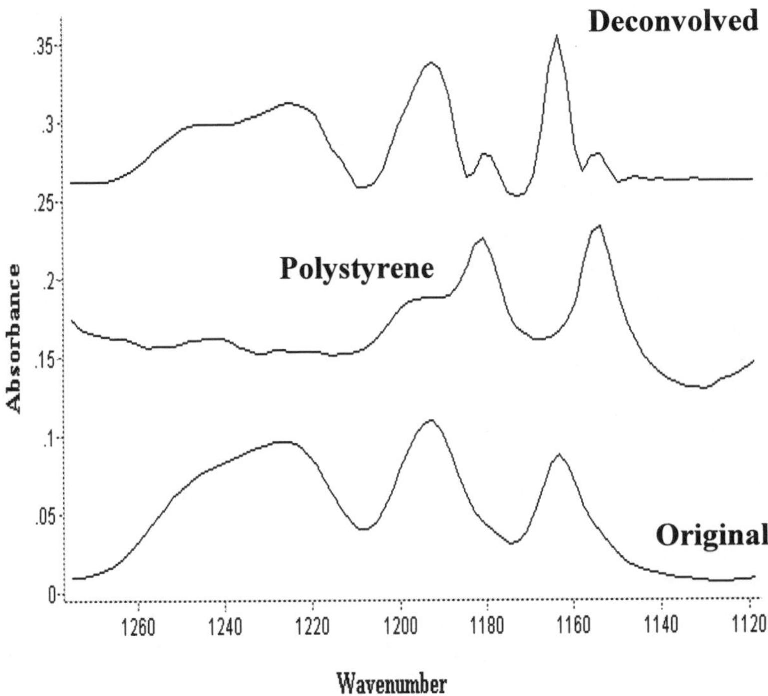

Figure 3.15 Bottom: The original undeconvolved spectrum of a polycarbonate/ polystyrene mixture. Middle: The spectrum of pure polystyrene. Top: The deconvolved spectrum of the mixture. Note that the features at 1180 and 1154 in the top spectrum line up exactly with the two polystyrene bands.

tives have three features in them corresponding to each absorbance band in the original spectrum. The number of downward pointing features in the second derivative is equal to the number of absorbance bands in the original spectrum. The lowest point of a downward pointing feature in a second derivative corresponds to the wavenumber of peak absorbance in the deconvolved spectrum. This is well illustrated in Figure 3.14. The original spectrum is at the bottom, the middle spectrum has been deconvolved, and the top spectrum is the second derivative of the original spectrum. Visual inspection of the original spectrum does not indicate the presence of six bands in this region. The second derivative clearly shows this, and it is confirmed by the results of the deconvolution. The spectrum was deconvolved until these six features were observed.

The spectrum shown in Figure 3.14 is of a mixture of a polystyrene and a polycarbonate. Before deconvolution the region shown consisted of four bands due to the polycarbonate, but no apparent bands due to the polystyrene. After deconvolution there are six bands observed. A comparison of a pure polystyrene spectrum to the deconvolution results is seen in Figure 3.15. It shows that the bands at 1180 and 1154 cm^{-1} in the deconvolution result are due to the polystyrene. As a result of the deconvolution, bands from different components that overlap have been resolved. This indicates that deconvolution has utility in trying to resolve overlapped bands in mixture spectra.

A feature of the deconvolution process is that in a group of overlapped bands, the narrower bands appear first as resolution is enhanced. For example, water vapor lines are very narrow, and if the band being deconvolved contains water vapor lines, these will appear in the deconvolved spectrum before all other features. This makes sense since narrow bands need less resolution enhancement to show up than broad bands. When interpreting the results of deconvolutions performed on regions where atmospheric gases absorb, make sure the features you see are not due to carbon dioxide or water vapor.

When the cepstrum is multiplied by the exponential function, the cepstrum is truncated in a similar fashion to how interferograms are truncated before they are Fourier transformed. As a result of truncating an interferogram, sidelobes are introduced into the spectrum, and these are eliminated by multiplying the interferogram by an apodization function (see Chapter 2). The same phenomenon holds true for cepstrums; they need to be multiplied by an apodization function prior to Fourier transformation to eliminate the "feet" in the deconvolved spectrum. There are several choices of apodization function that can be used in deconvolution. In reading the literature and through practical experience, the Bessel apodization function seems to work best [2].

Deconvolution can be a tricky spectral manipulation technique to use properly. However, if the guidelines discussed in this section are followed, deconvolution can be a powerful tool for revealing the structure of overlapped bands.

G. Spectral Library Searching

The main use of library searching is as an interpretation aid. The idea is to compare an unknown (or sample) spectrum against a collection of spectra of known compounds contained in a library. It is assumed that if two spectra are similar, the molecules in the two samples are similar. A library match is called a "hit". A good hit occurs when the match between two spectra is close, a bad hit occurs when two spectra are very dissimilar.

Before we discuss the attributes of library searching, it is important to realize what the results of a library search mean. The computer will always find a match for your unknown amongst the spectra searched, even if it is a lousy match. Always visually compare the unknown spectrum to the spectra of the library hits. Often times the computer may indicate two spectra are well matched, yet you may be able to see vast spectral differences between them. In these cases, trust your own judgement, use your own knowledge of the sample, of spectroscopy, and of spectral interpretation to guide you in interpreting search results. For instance, if you know there is not a ketone in your sample, yet the best library hit is a ketone, ignore the search results. Library searching is not a substitute for good science or interpretive skills; it is a tool to enhance the skills of the analyst.

There are several different types of library searches that can be performed, and these are a function of the software package you use. Firstly, one can perform a *text search.* Information about a sample whose spectrum appears in a library, such as its chemical names, physical properties, and chemical structure are stored in the library along side the sample's spectrum. In a text search, a string of text input by the user is searched against the textual information in a library. For example, if you know your sample is a polystyrene copolymer, you may want to perform a text search using the word "polystyrene" to narrow potential hits to only those samples that contain polystyrene. The second type of search that is called a *peak table search.* In this type of search, the peaks in the unknown spectrum are analyzed and placed in a peak table, and the spectrum is said to be "peak picked". Each library spectrum is peak picked, and the peak tables of the unknown and library spectra are compared. Whether or not two spectra are similar is based on how similar their peak tables are to each other. The problem with peak table searches is that only a peaks in the spectrum are used, so many potentially useful peaks are ignored. The best search method is to use what is known as a *full spectrum search.* This method the entire unknown spectrum compares it to each library spectrum in its entirety. A typical infrared spectrum contains thousands of data points, and using all of them in a library search greatly increases the odds of obtaining an accurate search result. In essence, if the data points are available, they should be used in the library search.

The first question that needs to be answered about library searching is where to obtain spectral libraries. In the past, spectral libraries were only available in book form. Today, spectral libraries come in digitized form on floppy disks, and are simply installed onto your computer's hard disk. Many FTIR companies sell spectral libraries, but their collections tend to be small or specialized. There are two major commercial sources of spectral libraries. The company with the largest collection of spectra is the Sadtler Division of Bio-Rad (Philadelphia, PA). They have over 150,000 digitized spectra of almost every kind of compound. Fortunately you don't have to buy all 150,000 spectra; smaller specialized libraries of spectra of samples related to each other can be purchased. For example, there are polymer, inorganics, gas phase, and pollutants libraries. Sadtler libraries can be used with the search software sold by most FTIR instrument manufacturers. The second commercial source of infrared libraries is Aldrich Chemical Company (Milwaukee, WI). They have about 50,000 spectra in their collection, and also have smaller specialized libraries available. Regardless of who you buy your library from, it is a good idea to have a collection of commercial libraries containing spectra of sample types you analyze frequently.

Despite the availability of commercial infrared libraries, one of the best sources of libraries is you, the user of an FTIR. Making your own libraries is

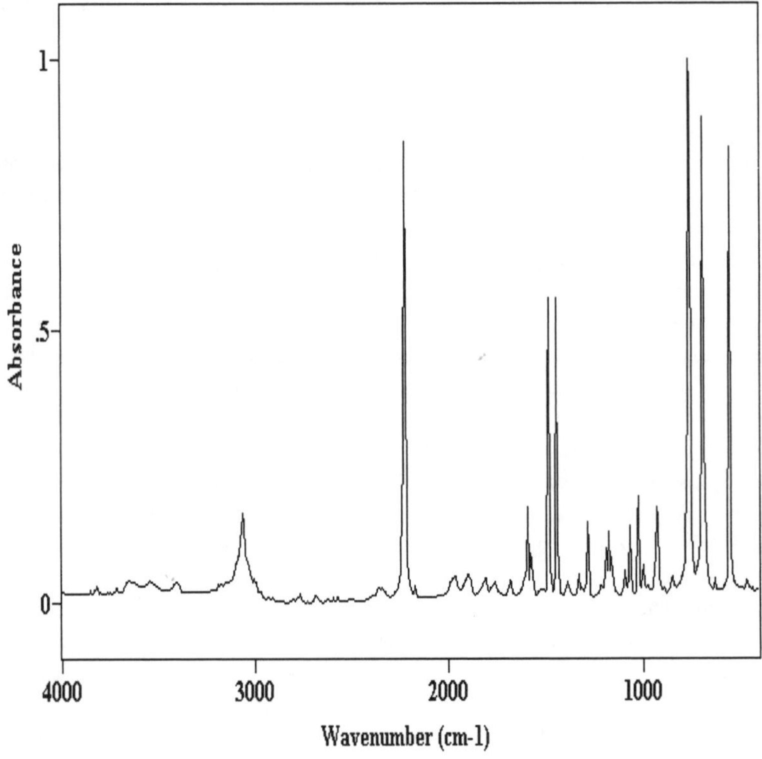

Figure 3.16 An example of a normalized spectrum. Note the intensity scale is from zero to one.

a good idea because only you have access to the kinds of samples that are typical in your work. For example, if you identify an unknown today and put it in a library, you will always have that information at your disposal. If you encounter the same sample 3 years from now, you will be able to quickly and easily identify it since it will show up as a good hit when the unknown is searched. When making your own libraries, spectra of samples that are related to each other are best kept together. For instance, a user generated library could contain the spectra of all the raw materials that a quality control lab analyzes, spectra of all the batches of a finished product, or all the polymers ever synthesized by an organic chemistry group. All the spectra in a library must be of the same resolution, cover the same wavenumber range. Library spectra are used as references, so the spectra you place in a library should have a high SNR and be free of artifacts. Also, spectra added to a library should be in absorbance units (the software may make the conversion from transmittance for you). Spectra added to a library are usually baseline corrected. Baseline slope is usually caused by instrument or sampling problems, and it is undesirable to have these contribute to search results.

Library and unknown spectra are often times *normalized.* Normalization means all the absorbances in a spectrum are divided by the largest absorbance in a spectrum. For example, if the biggest peak in a spectrum has an absorbance of 0.8, all the absorbances in the spectrum are divided by 0.8. The intensity scale for all normalized spectra is from 0 minimum to 1 maximum. A normalized spectrum is seen in Figure 3.16. The purpose of normalization is to remove differences in peak heights between spectra acquired under different conditions (i.e., different concentrations and pathlengths). Essentially, it allows an apples-to-apples comparison to be made [3]. Since spectra added to a library are altered, it is highly recommended that a copy of the unaltered spectrum be retained.

Once you have an unknown spectrum that you want to search, it is important to use the right libraries. Unknown spectra should be searched against libraries that contain spectra of samples similar to the unknown. For instance, if you search a gas phase spectrum against a polymer library, you will not get good results because polymers do not exist in the gas phase! However, you may want to search a polymer spectrum against a polymer library and a general organics library, since many polymers are organic materials. If your first library choice does not yield good results, be creative and try some other libraries.

Once the unknown spectrum and the library to search have been chosen, the software takes over and the search process begins. In the first step of the process, the unknown spectrum is converted to absorbance (if needed), normalized, and baseline corrected. Next, the unknown spectrum is compared to each library spectrum using a *search algorithm.* The comparison produces a number called the *hit quality index* (HQI), which is a direct measure of the similarity between two spectra. There are many different types of search algorithms, as will be discussed below.

The next thing the search software does is rank the library spectra by their HQI, from best to worst. This list is reported to the user as a *search report.* The search report often contains the name, index number, library name, HQI, and other important information about the library spectrum. Figure 3.17 shows a search report, and a comparison of an unknown and a library spectrum. A spectrum of chloroform was used in the search. Note that these two spectra look almost identical, and that the HQI for the best hit is 0.031, where 0.0 is a perfect hit. This search was successful since the best hit is the same as the "unknown" sample. As mentioned above, you should always visually compare the unknown spectrum to the library hits to insure that the search results are accurate.

The hit quality index is probably the most important part of a search report. Unfortunately, many different systems are used by different search software packages to calculate the hit quality index. Some systems call 0.0 a perfect match, and 1.414 (the square root of 2) the worst possible match. Other sys-

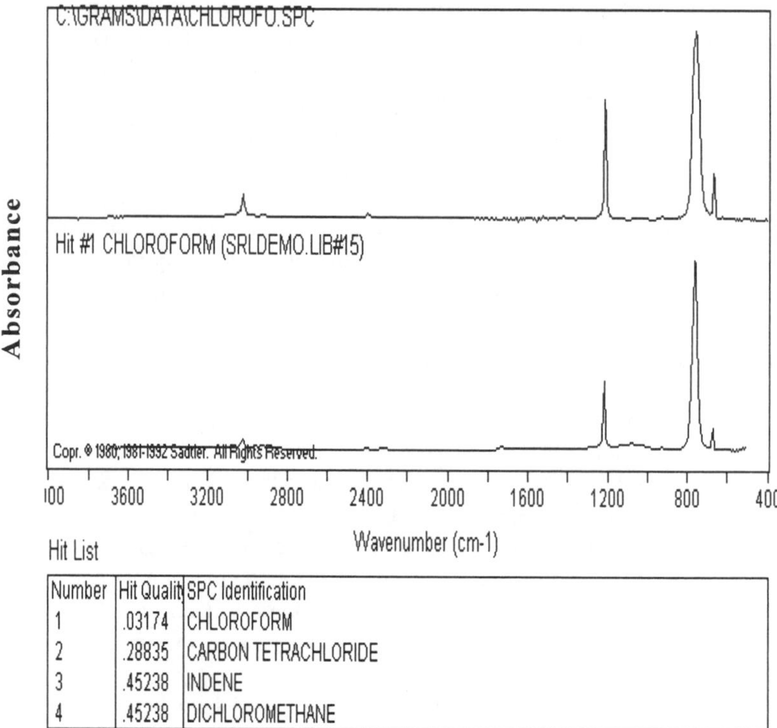

Number	Hit Qualit	SPC Identification
1	.03174	CHLOROFORM
2	.28835	CARBON TETRACHLORIDE
3	.45238	INDENE
4	.45238	DICHLOROMETHANE

Figure 3.17 The results of a library search of a chloroform spectrum. The unknown and library spectra are at the top. The bottom shows part of the search report. In this system, an HQI of 0.0 is a perfect match.

tems use 1, 10, and 1000 as the HQI for a best match. This makes it very difficult to compare searches performed with different FTIR software packages. To get around this problem, a system has been developed to normalize the different hit quality indices obtained with different software packages. In this system, hit quality indices are divided into 3 ranges corresponding to a 0 to 20% difference between two spectra, a 20 to 50% difference, and a greater than 50% difference. These ranges have been dubbed the "close", "suggestive", and "be careful" ranges, respectively. Table 3.2 shows the HQI values that correspond to these ranges for several different software packages. If two spectra are less than 20% different the match is in the close range, and means either a perfect match has been found or the two spectra are very similar to each other. Sometimes, if the HQI is in this range, a complete molecular identification can be achieved as seen in the search report in Figure 3.17.

If two spectra are 20% to 50% different, the HQI will fall in the suggestive

Table 3.2
The Close, Suggestive, and Be Careful HQI Ranges

HQI Limits	Close	Suggestive	Be Careful
0.0 - 1.41	<0.3	0.3 - 0.7	>0.7
0.0 - 1.0	<0.2	0.2 - 0.5	>0.5
1.0 - 0.0	>0.8	0.5 - 0.8	<0.5
10.0 - 0.0	>8.0	5.0 - 8.0	<5.0
1000 - 0	>800	500 - 800	<500

range. This means there are some significant differences between the spectra, and a complete molecular identification is impossible. However, there are enough similarities between the spectra to suggest what functional groups may be present in the unknown, or what class of chemical compounds the unknown belongs. Finally, if the difference between two spectra is greater than 50% , the HQI will fall into the "be careful" range. This means there is little or no similarity between the unknown and the library spectrum. Great care must be taken when interpreting the HQIs in this range. If upon visual examination of the spectra there are some similarities between the unknown and the library spectrum, you are in luck. However, if there is little or no similarity between the two spectra, ignore the search results.

If you don't find the HQI limits of your software represented in Table 3.2, you can calculate the three HQI ranges for your software package as follows. Take the number for a perfect hit and multiply by 0.80, this is the lower limit for the close range. Multiply the perfect hit number times 0.5 to find the lowest HQI in the suggestive range. All HQIs below the suggestive range are in the be careful range.

Another thing to look at in the search report is the relative difference between HQIs. This is calculated by taking the difference in HQI for two consecutive hits, and dividing the difference by the smaller HQI. For example, in the search report in Figure 3.17, the relative difference in HQI for hits 1 and 2 is (0.031 - 0.29)/(0.031) or 835%. If the difference between consecutive HQIs is greater than 10%, then all hits below this point in the search report can normally be ignored. For the search report in Figure 3.17, the first break in consecutive HQIs greater than 10% takes place after the first hit. All other hits in this search report should be ignored. This confirms that the library search completely identified this particular molecule.

Here is how one simple search algorithm, the absolute value algorithm, works. First, each library spectrum is subtracted from the sample spectrum. The result of this calculation is called a residual. An example of a residual is seen in Figure 3.18. Note that it has upward and downward pointing features, and looks like any other subtraction result. The size of the residual is directly related to how similar two spectra are to each other. Spectra that are identical

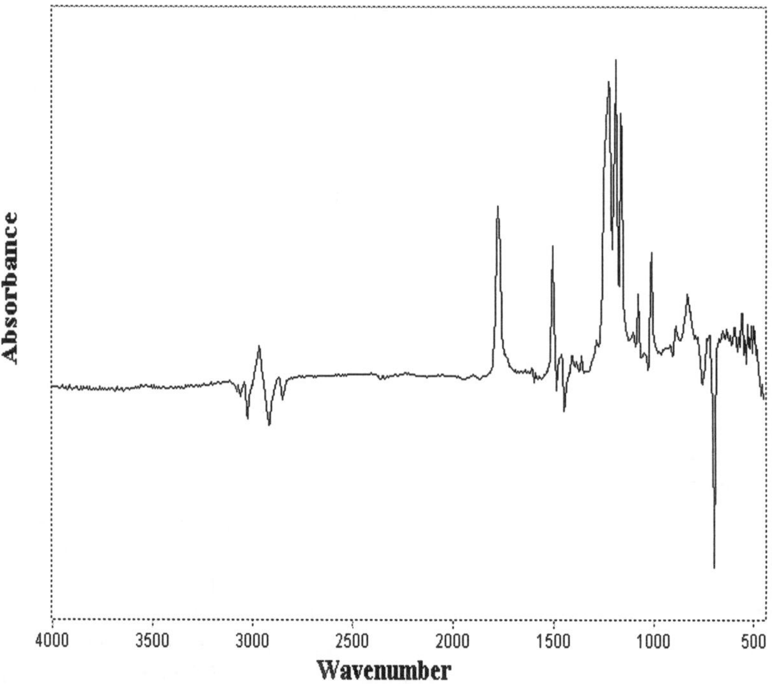

Figure 3.17 A residual calculated by subtracting a library spectrum from an unknown spectrum. Note the presence of positive and negative pointing features.

will have a residual of zero (a straight line), spectra that are very dissimilar will have large residuals. The search program measures the size of the residual by taking its absolute value and dividing by the number of data points. This calculation gives the HQI.

The search algorithms available to you depend on your search software package. The correlation algorithm is probably the most common search algorithm used, and here is how it works. First, this algorithm calculates the average absorbance of the unknown spectrum and of each library spectrum, then the appropriate absorbance is subtracted from each spectrum. Then the algorithm treats the spectra as vectors and calculates the dot product of the unknown and library spectra to obtain an HQI. The Euclidean distance algorithm is also widely used. It uses a dot product like the correlation algorithm, but the spectra are normalized rather than having the average absorbance subtracted out. If the search algorithm contains the words least squares in its title, the library and unknown spectra are subtracted to obtain a residual, then the square of the residual is used to calculate the HQI. Any search algorithm that

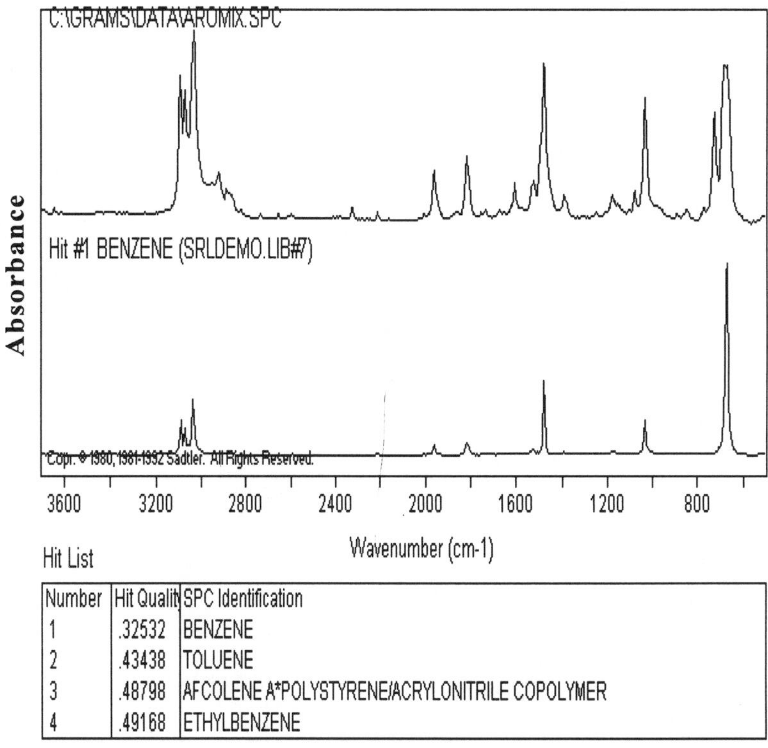

Number	Hit Qualit	SPC Identification
1	.32532	BENZENE
2	.43438	TOLUENE
3	.48798	AFCOLENE A*POLYSTYRENE/ACRYLONITRILE COPOLYMER
4	.49168	ETHYLBENZENE

Figure 3.19 The search results for a mixture of benzene and toluene. The correlation algorithm was used, and the HQI for a perfect hit is 0.0.

has the word "derivative" in it's name compares the derivative of the unknown spectrum to the derivatives of the library spectra. Derivative algorithms are useful for spectra with baseline slope, since spectral derivatives have a flat baseline. The different search algorithms also differ in how they weigh the importance of peak position and relative intensity in matching spectra. The correlation and Euclidean distance algorithms put equal emphasis on intensity and position, and are used for a wide variety of spectra. Algorithms that use derivatives tend to emphasize peak position over peak intensity. These are very useful when searching gas phase and GC-FTIR spectra because these samples have many narrow, well defined bands. Lastly, any algorithm that uses absolute values tends to emphasize relative intensity differences over peak positions.

In addition to determining the identity of a pure unknown, library searching has some utility in identifying components in mixtures. If one is searching the spectrum of a mixture, and one of the major components in the unknown

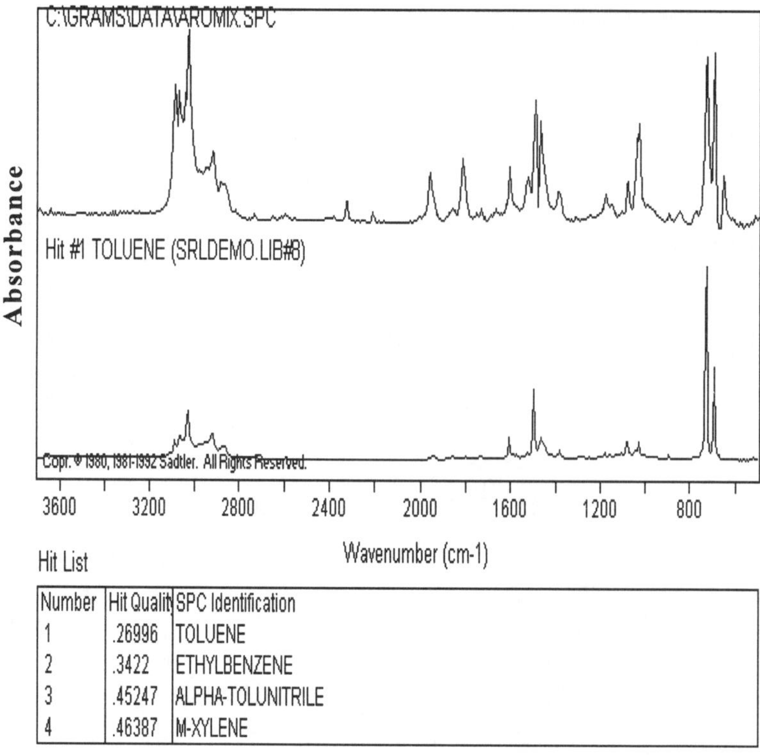

Number	Hit Qualit	SPC Identification
1	.26996	TOLUENE
2	.3422	ETHYLBENZENE
3	.45247	ALPHA-TOLUNITRILE
4	.46387	M-XYLENE

Figure 3.20 An example of a subtract and search again. The benzene spectrum seen in Figure 3.19 was subtracted from the spectrum of a benzene/ toluene mixture. The subtraction result was searched again, and the second component in the mixture was correctly identified as toluene. The correlation algorithm was used, and 0.0 is the HQI for a perfect hit.

turns up as a library hit, you can subtract the component's spectrum from the unknown and search again. With luck, the spectrum of a second component will be a top hit, which means the identity of the second component can be obtained. This is known as the *subtract and search again* process, and is illustrated in Figures 3.19 and 3.20. The spectrum of a mixture of benzene and toluene was obtained and searched against a library that contains spectra of both pure components. The results of the search of the mixture are shown in Figure 3.19. The correlation algorithm was used, and an HQI of 0.0 denotes a perfect hit. The best hit is the spectrum of benzene. The benzene library spectrum was subtracted from the mixture spectrum, and the result was searched against the same library as used above. The results of this search are seen in Figure 3.20 (using same HQI system and algorithm as above). The best hit is toluene, so these library searches correctly identified both components in the mixture. Now, for complex mixtures your results may not be as clear cut as this simple example. From a practical viewpoint, successfully searching and subtracting one spectrum more than two or three times in a row is rare.

H. Curvefitting

Curvefitting addresses the same problem as deconvolution, trying to determine the position and intensity of a several peaks that overlap to give a broad spectral feature. However, curvefitting attacks the problem in a totally different fashion than deconvolution as follows. The broad, overlapped feature of interest is examined, and an estimate of the number, width, height, location, and shape of the underlying peaks is needed. To assist in this process, the second derivative of the spectrum can be examined, or the spectrum can be deconvolved prior to curvefitting (see above for a discussion of derivatives and deconvolution). Based on this assessment of the spectrum, an initial estimate as to the nature of the underlying peaks is entered into the computer.

An example of an initial guess for a curvefit is seen in Figure 3.21. The top spectrum is of a polystyrene/polycarbonate mixture, and is the same spectrum whose deconvolution results are seen in Figure 3.14. The middle spectrum in the figure shows the four peaks to be used in the initial estimate, and the bottom shows the residual. The residual is the difference between the actual spectrum and the spectrum calculated using the current set of curvefitting parameters.

Once the initial estimate has been entered into the computer, the software uses a least squares fitting algorithm to vary and optimize all the parameters. Using this algorithm guarantees that the best possible fit will be found. Curvefitting is an iterative process, and the program may need to run many times to achieve good results. The first guess is used in the first iteration, and the results of the previous iteration are used as the initial guess for the next iteration. The quality of the fit is determined by comparing the calculated spectrum to the real spectrum. A number, usually a standard deviation, a standard error, or a "chi-square" is used for this purpose. With some software packages, it is up to the user to monitor the quality of the fit, and to stop the process once the fit quality is optimized. Otherwise, the software will monitor the quality of the fit and stop the process at an appropriate point.

The results of the curvefitting process are a set of bands with specific widths, heights, positions, and shapes. When combined, these bands should generate a spectrum that closely matches the sample spectrum. The result of a curvefit is seen in Figure 3.22. This fit resulted from the initial guess seen in Figure 3.21. The top of the figure shows the sample and calculated spectra, the middle shows the bands that comprise the calculated spectrum, and the bottom shows the residual, which is the difference between the calculated and original spectra. Note that the residual is smaller in Figure 3.22 than in Figure 3.21, which indicates that the fitting process did improve the agreement between the real

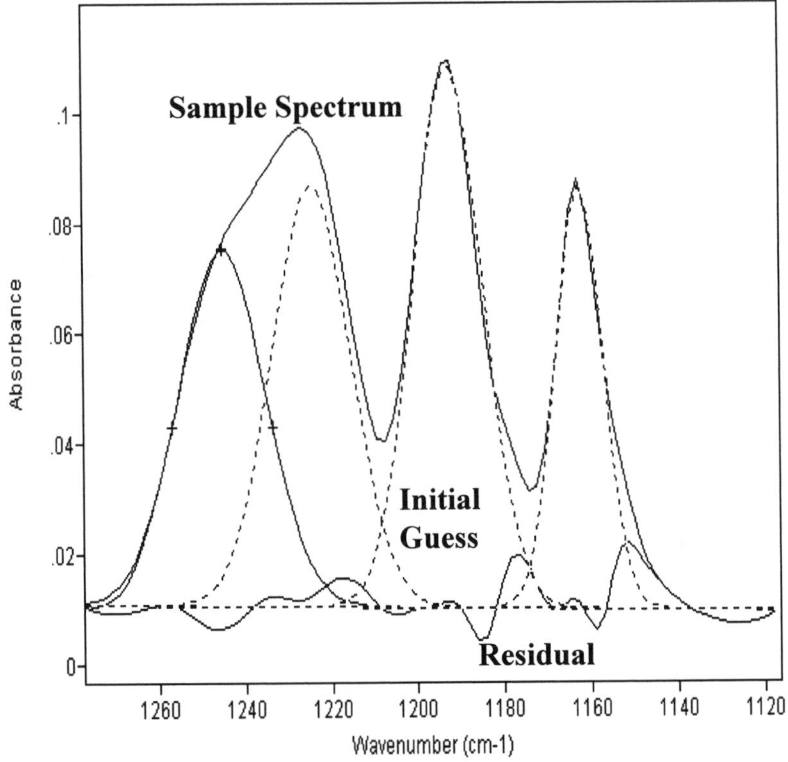

Figure 3.21 An example of an intitial a guess at a set of bands to be used in a curvefit. The top spectrum is of the sample, the middle displays the bands to be used as the initial guess, and the bottom shows the residual.

and calculated spectra. Once a curvefit has been successfully completed, it is possible that the curvefitting parameters describe the true infrared spectrum. If this assumption is made, then conclusions about the sample can be made. For instance, based on the results shown in Figure 3.22 you may conclude that the spectrum that was used in the calculation has four bands in the region studied. However, one must be careful in interpreting curvefitting results. Make sure the quality of the fit is a good one; or you may make false conclusions.

The piece of spectrum used in the fit and shown in Figure 3.22 is comprised of six bands, as seen in the second derivative and deconvolution results presented earlier in this chapter. In reality, six peaks should have been used in the curvefit. The reason only four peaks were used in the fit is to illustrate that although a set of parameters may give a good fit, they may still not be the best or most accurate set that is possible.

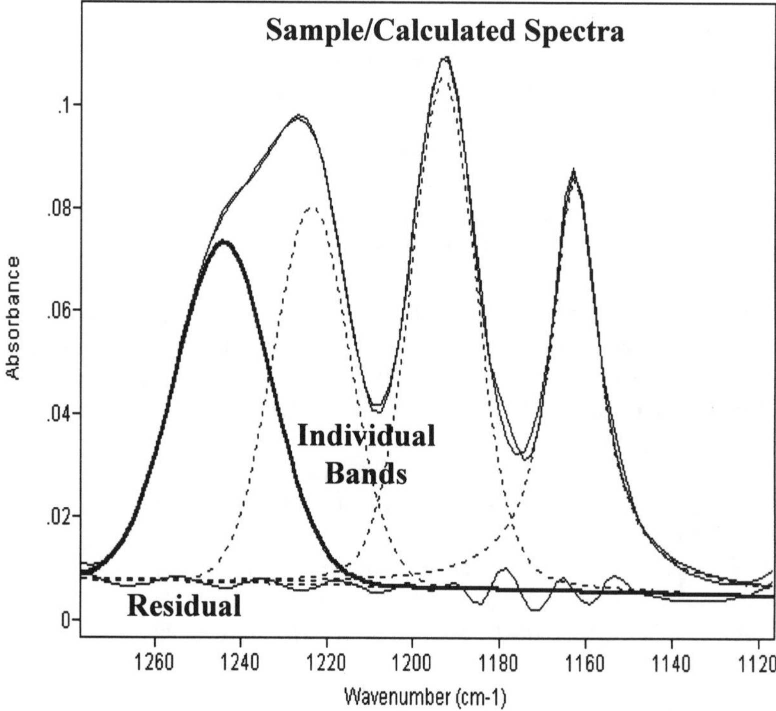

Figure 3.22 The results of a curvefit. The sample spectrum and the calculated spectrum are almost perfect overlaps as seen in the top of the display. The middle shows the bands that make up the calculated spectrum, and the bottom shows the residual.

The biggest problem with curvefitting is not knowing whether the calculated set of parameters is <u>unique</u>. It is entirely possible that several sets of curvefitting parameters will give equally good results. In this case it is not obvious which set of parameters to believe. The deconvolution method discussed above does not suffer from a uniqueness problem (although it is not always obvious what resolution enhancement factor to use). When trying to determine the structure underlying a broad, overlapped band, the author prefers to use deconvolution first. This is because deconvolution is generally simpler than curvefitting, and it does not suffer from a uniqueness problem.

I. Conclusion

It has been shown how the many different kinds of spectral manipulation can enhance the appearance and information content of a spectrum. However, since spectral manipulations always involve altering the original data generated by the instrument, the data may be changed in a way that causes misinterpretation. Therefore, it is critical that you follow the Laws of Spectral Manipulation as put forth above, and pay heed to the advice given in each section of the chapter on how to avoid data destruction.

References

1. A. Savitsky and A. Golay, *Anal. Chem.* **36**(1964)1627.
2. J. Kaupinen, D. Moffat, H, Mantsch, and D. Cameron, *Appl. Spec* **35**(1981)271.
3. S. Standford, *Annals of Improbable Research* **1**(3)(1995)2. Much has been made of not comparing "apples to oranges". This paper shows, through recent infrared spectroscopic work, that apples and oranges really <u>are</u> the same, at least from the point of view of an infrared spectrometer.

Bibliography

P. Griffiths and J. de Haseth, *Fourier Transform Infrared Spectroscopy*, Wiley, New York, 1986.
W. George and H. Willies, Eds., *Computer Methods in UV, Visible, and IR Spectroscopy*, Royal Society of Chemistry, Cambridge U.K., 1990.
S. Johnston, *Fourier Transform Infrared: A Constantly Evolving Technology*, Ellis Horwood, London, 1991.

Chapter 4

Choosing the Right Sampling Technique

The purpose of this chapter is to discuss the most commonly used infrared sampling techniques. Each technique is intended for use with specific types of samples, and has its own strengths and weaknesses. Therefore, the samples for which each technique is well suited will be emphasized. Appropriate theory will be discussed along with "tricks of the trade" that will make your sample handling simpler and easier.

A. Transmission Techniques

Probably the most popular way of obtaining infrared spectra is to pass the infrared beam directly through the sample as shown in Figure 4.1. This is known as the *transmission technique* of sampling. The advantages of this technique are that transmission spectra have high signal-to-noise ratios, and comparatively inexpensive tools are used to prepare the samples for this type of analysis. Another advantage of transmission sampling is that it is a universal technique: it works on solids, liquids, gases, and polymers.

One of the major disadvantages of transmission techniques is the "thickness problem". Generally, samples thicker than 20 microns absorb too much infrared radiation, making it impossible to obtain a spectrum. Samples thinner than 1 micron have absorbances too weak to be detected by the spectrometer. Ideally, samples should be between 1 and 20 microns thick. The challenge in preparing transmission samples is to adjust the thickness or concentration of samples so the appropriate amount of light passes through the sample. Another disadvantage of transmission techniques is that time consuming sample preparation may be required. It takes time to melt, squish, or dilute a sample enough so that it transmits the appropriate amount of light. The following discussion of transmission techniques is organized by the type of sample being analyzed.

A.1. Transmission Spectra of Solids

The transmission sampling techniques for solids can be divided into two

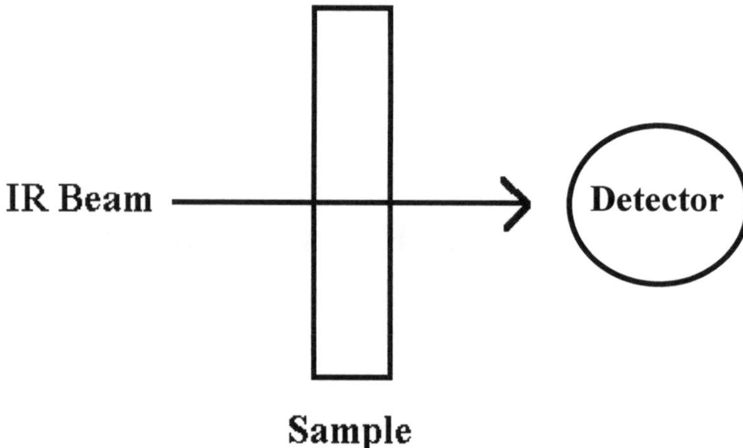

Figure 4.1 An illustration of transmission FTIR sampling, where the infrared beam passes through the sample then strikes the detector.

classes. First, the techniques best suited for powders or samples than can be ground into a powder are called KBr pellets and mulls. These methods focus on grinding the sample to reduce its particle size, then diluting the sample in an inert matrix so the appropriate amount of light passes through the sample. The second class of transmission sampling methods for solids involves two ways of preparing polymeric samples as thin films for infrared analysis. The goal in preparing a polymeric film is to make the film between 1 and 20 microns in thickness so it lets the right amount of light through. Powders are incapable of forming films, so cannot be analyzed as films.

KBr Pellets

Potassium bromide (KBr) pellets are used to obtain the infrared spectra of solids, and are particularly well suited to powdered samples. KBr is an inert, infrared transparent material, and acts as a support and a diluent for the sample. Here are the steps to follow for preparing successful KBr pellets. First, the sample and the KBr must be ground to reduce the particle size to less than 2 microns in diameter. Bigger particles will scatter the infrared beam and cause a sloping baseline as seen in Figure 4.2. Grinding is traditionally performed with an agate mortar and pestle, but a Wig-L-Bug^{TM} may also be used (see below). A gram or so of KBr should be placed in the mortar. It should be ground until crystallites can no longer be seen and it becomes somewhat "pasty" and sticks to the sides of the mortar. The KBr and the sample should be ground separately to avoid possible chemical interactions; the heat and pressure generated in the mortar may cause the KBr to react with the sample. The

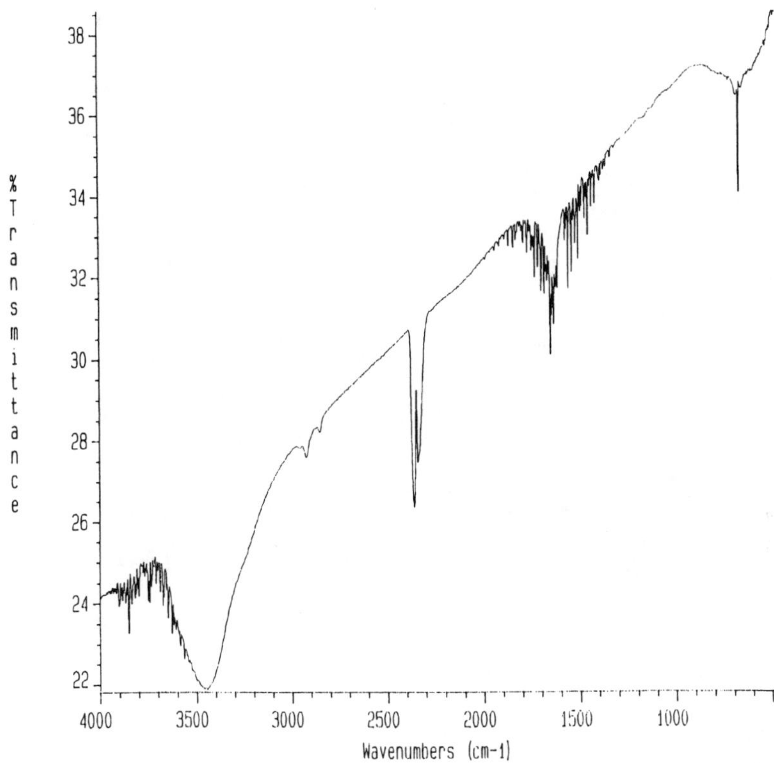

Figure 4.2 An example of a bad KBr pellet spectrum. The sloping baseline is caused by the particle size being too large. The water peaks at 3450 and 1630 cm^{-1} are caused by using wet KBr.

spectrum obtained may be that of the product of this reaction rather than of your original sample. After grinding the sample and KBr, the sample is diluted to about 1% in the ground KBr. About 1 to 10 mg of sample can be used. The amounts of material to use can be "eyeballed" with some success, but actually weighing out the amounts will give more reproducible results. It is important that the sample be well dispersed in the KBr. Mixing the sample and KBr in the mortar for a minute using a spatula is usually sufficient to ensure good dispersion.

The sample/KBr mixture is then placed in a dye or press, and is squeezed to produce a transparent pellet. Several tons of pressure may be necessary to obtain a transparent pellet. If the pellet is inside a dye, place the pellet/dye combination directly in the infrared beam. If the pellet is freestanding, it should be placed in a pellet holder then placed in the IR beam. The background spectrum should be obtained on the empty dye or pellet holder. The

Figure 4.3 A KBr "mini-pellet maker". (Photo courtesy of Spectra-Tech Inc.)

background spectrum should not be obtained on a KBr pellet made without sample. Such a pellet would need to have optical properties identical to those of the sample pellet to be useful as a background. Since this is impossible in practice, the empty pellet holder should be used for the background spectrum.

Figure 4.3 shows a picture of an inexpensive and easy to use KBr pellet press known as a "mini-press" or "mini-pellet maker". It consists of a stainless steel barrel and two stainless steel bolts. One bolt is screwed about half way into the barrel. The barrel is held upright, and enough KBr/sample mixture is added to cover the face of the bolt. The second bolt is screwed into the barrel while still holding the device upright. Next, the bolts are tightened using wrenches to apply the pressure needed to make a transparent pellet. After a minute or so, the barrel is held horizontally and the bolts are gently unscrewed. Hopefully, a thin, semitransparent KBr disk will be visible. If the pellet looks good, place the barrel in the infrared beam, and obtain the spectrum of the pellet.

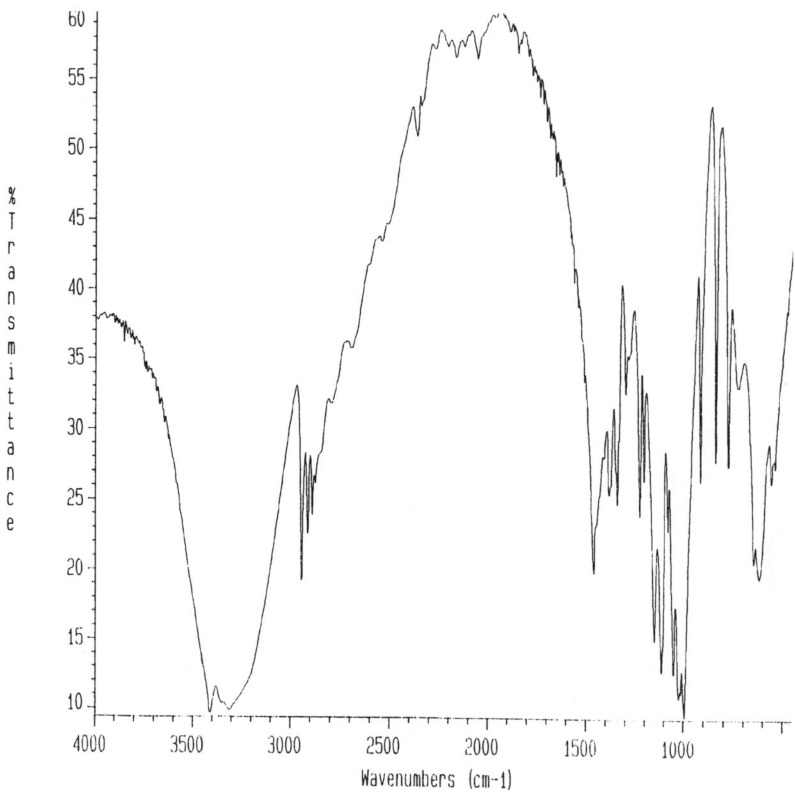

Figure 4.4 An example of a quality KBr pellet spectrum. This is the infrared spectrum of glucose.

There are several problems that may be encountered when making a KBr pellet. An opaque pellet indicates that too much material was used in making the pellet. Opaque pellets will give poor spectra because very little light will pass through them. White spots in a pellet indicate the sample was not ground well enough, or was not dispersed properly in the KBr. Try grinding a little longer and mixing the sample and KBr more thoroughly. If too little material is placed into the KBr pellet press, the result will be no pellet at all, just little piles of powder. The pressure applied to make the pellet also needs to be controlled, and must be high enough to produce the thin transparent film desired. Finally, handling the pellet can be a problem. Pellets are thin and brittle, and can fall out of pellet holders, crack, or disintegrate with the slightest provocation. Care must be exercised in handling them. A good KBr pellet is thin and transparent. A quality KBr pellet spectrum of glucose is shown in Figure 4.4.

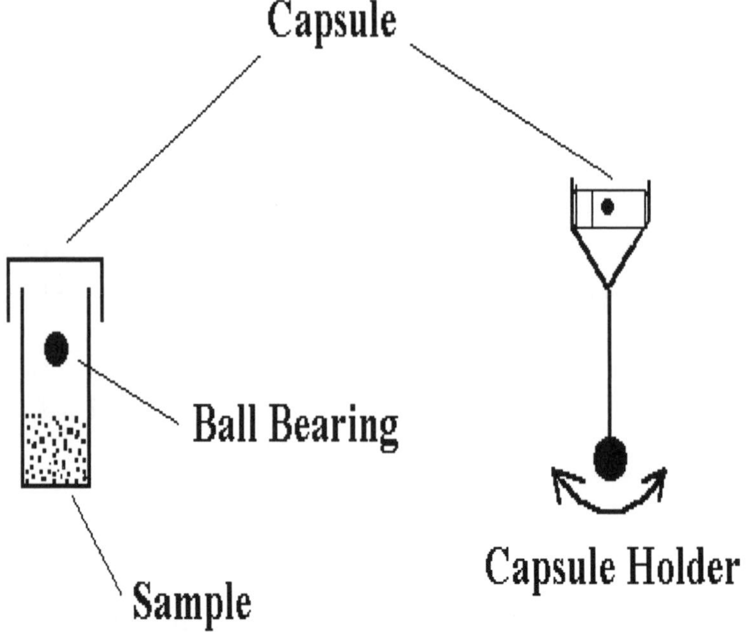

Figure 4.5 Diagram of a Wig-L-Bug, used to grind KBr and samples for use in KBr pellets. The material to be ground and the ball bearing are placed in a stainless steel capsule. The capsule is placed in a holder, which is shaken vigourously for 1 or more minutes to grind the material.

KBr is a hygroscopic material, which means it will absorb water directly from the atmosphere. It is critical that the KBr used in making pellets be kept warm and dry, preferably in an oven at >100° C. Cloudy regions in a pellet indicate the pellet has absorbed water, which will give bands around 3500 and 1630 cm^{-1} as seen in Figure 4.2. A freshly made pellet should not be left exposed to the air too long, for it will also absorb water.

Some samples are easily made into KBr pellets, whereas others do not give good pellets and an entire afternoon can be wasted trying to make a pellet. It is not obvious by looking at a sample whether it can be easily made into a pellet. This is where trial and error, a little luck, and patience go a long way towards the making a good KBr pellet.

One last word on grinding. For most purposes, an agate mortar and pestle are fine, but a device called a Wig-L-Bug™ exists for grinding that is fast and easy to use. Wig-L-Bugs were invented for use in dentists' offices to make amalgams for fillings (which is where you might have seen them before) and are manufactured by Crescent Dental Manufacturing Co. of Chicago, IL. Many laboratory supply companies (e.g., Fisher Scientific, VWR) sell Wig-L-Bugs. A diagram of a Wig-L-Bug is shown in Figure 4.5. The KBr or sample is placed in a stainless steel capsule along with a ball bearing. The capsule is placed in a holder that is then shaken at high speed. The ball bearing flies around inside the capsule smashing the sample or KBr into pieces. A minute or two of shaking is all that is necessary to pulverize most samples. This is faster and easier than manual grinding, and also gives a more reproducible particle size and particle shape distribution. The KBr and sample should be ground using separate capsules, then appropriate amounts of sample and KBr should be placed in a single capsule without a ball bearing. Shaking the contents of this capsule for a minute will insure good dispersion of the sample in the KBr with little extra grinding taking place. After this, making a KBr pellet is the same as described above.

The use of KBr pellets is widespread. Practically any powdered solid can be analyzed using KBr pellets. If a solid is not a powder, its spectrum can be obtained if it can be broken into pieces and ground or pulverized. If only small amounts of sample are available, "micro" KBr pellet presses are available that make pellets as small as several millimeters in diameter. KBr pellets are also well suited for quantitative work. One should measure out precise amounts of KBr and sample to use in the pellet when doing quantitative work. Also, since the pathlength of the pellet is not known, band area ratios must be used to eliminate pathlength as a variable. It is also recommended that the method of internal standards be used (see Chapter 5). KBr pellets, despite some of the difficulties involved in making a good pellet, are perhaps still the most popular technique for obtaining the infrared spectra of solids.

Mulls

Mulls are another way of obtaining the transmission spectra of solid materials. The sample, usually a powder, is ground in an agate mortar and pestle to achieve a small particle size and avoid scattering. A drop or two of oil, called the “mulling agent” (hence the name of the technique) is added, and the mixture is ground some more to disperse the solid in the oil. The mulling agent is usually mineral oil, which is sold under the trade name Nujol. This technique is called the "Nujol mull" technique because of the prevalence of this brand of mineral oil. To obtain a spectrum, a small amount of the sample/oil slurry is smeared on a KBr window, and a second KBr window is squeezed against the first to make a thin film "sandwich". This is illustrated in Figure 4.6. The film and windows are then placed in the infrared beam. The resultant spectrum will be of the sample and the mulling agent. The background spectrum should be run using the same KBr windows that will be used in the mull, but with nothing between the windows.

The advantage of mulls over KBr pellets is that mulls are easier and faster to make since there is no time consuming pellet squeezing involved. Mulls are

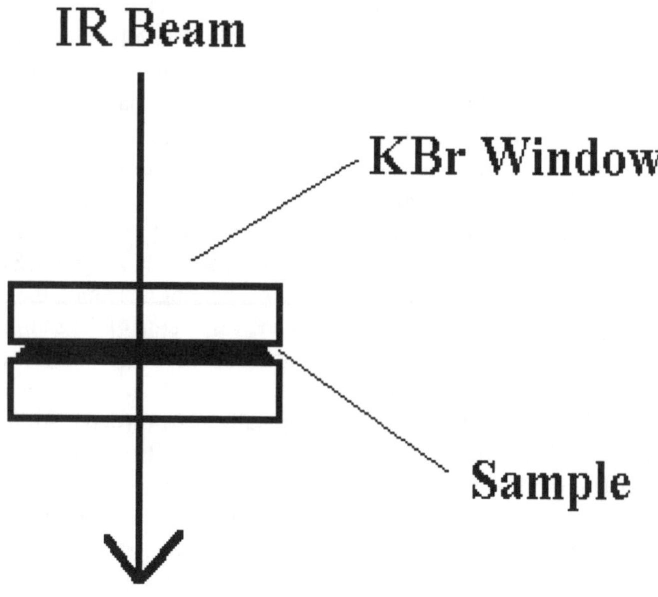

Figure 4.6 A diagram of a mull sample. This is sometimes called a "sandwich".

also well suited to samples that may degrade upon exposure to water. The mulling oil protects the sample from water vapor in the atmosphere since the oil is hydrophobic. KBr is hygroscopic and adsorbs water out of the atmosphere, potentially damaging water sensitive samples.

The disadvantage of the mull technique is that the infrared bands of the mulling agent contaminate the spectrum of the sample. The infrared spectrum of Nujol is shown in Figure 4.7. Nujol is a mixture of long, straight chain hydrocarbons, and it absorbs strongly around 3000 and 1400 cm^{-1}. In theory, one could subtract the spectrum of Nujol from the sample spectrum. However, the bands from the oil are usually too intense for clean subtractions to take place.

Fortunately, there is a way of getting around the problem of Nujol's infrared bands using a technique called the *split mull method.* This involves making up two mulls using different mulling agents. The first mull is prepared in Nujol, the second in an oil called Fluorolube. Fluorolube is a mixture of chlorofluorocarbons, long chain alkanes where the CH bonds have been replaced by CF and CCl bonds. The infrared spectrum of Fluorolube is shown

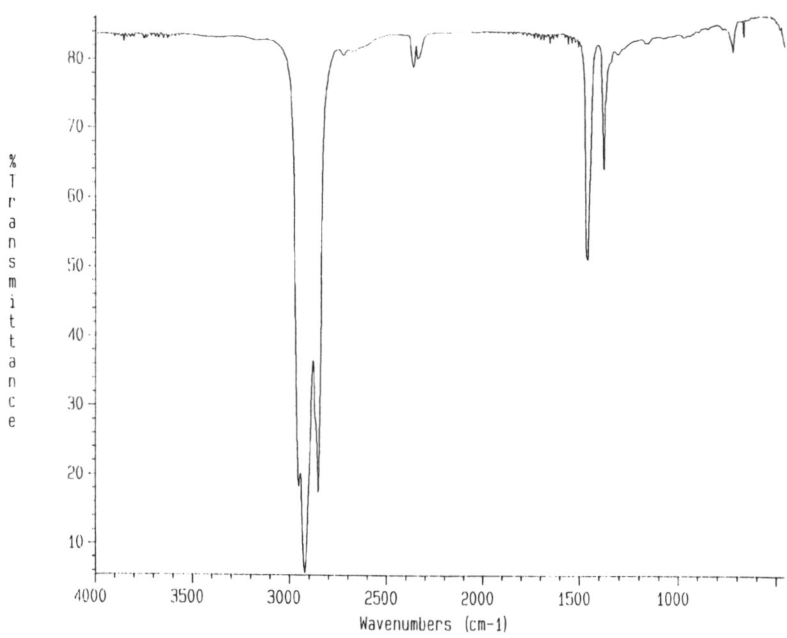

Figure 4.7 The infrared spectrum of Nujol, a hydrocarbon oil used as a mulling agent. Note the bands around 3000 and 1400 cm^{-1}.

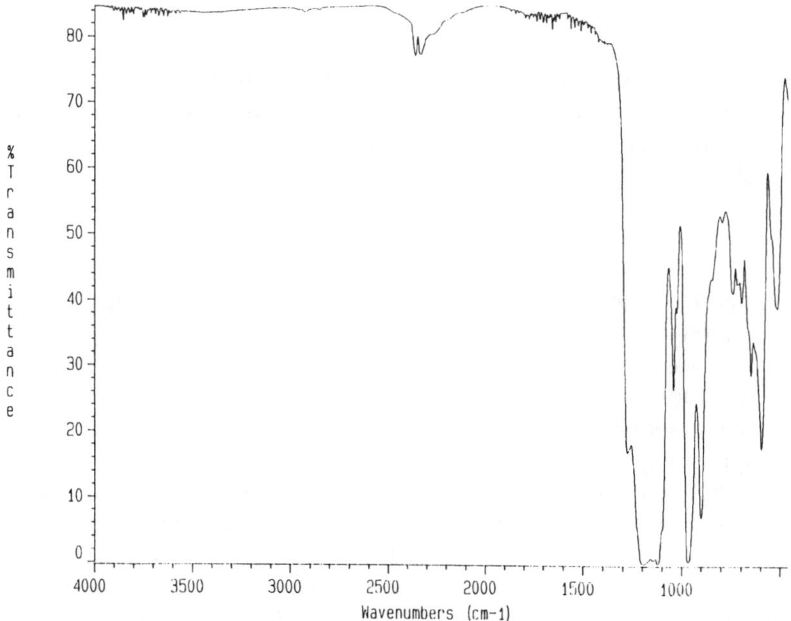

Figure 4.8 The infrared spectrum of Fluorolube, a mulling agent. Note the infrared bands below 1300 cm^{-1}.

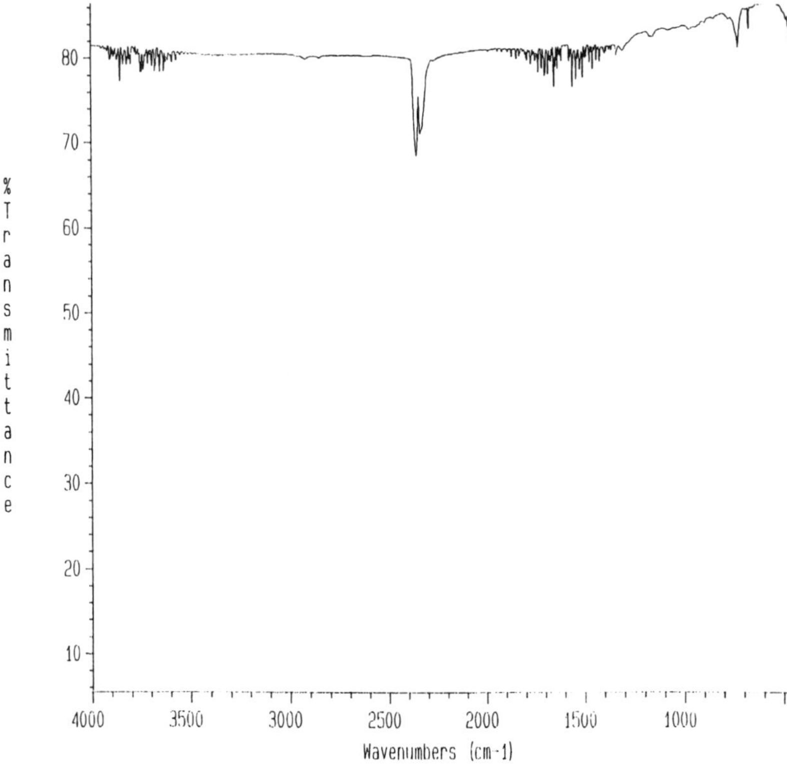

Figure 4.9 The spectrum of Nujol from 4000 to 1300 cm^{-1} spliced with the spectrum of Fluorolube from 1300 to 400 cm^{-1}. Note the absence of bands due to either oil (except for the Nujol peak at 720 cm^{-1}). This is the basis for the split mull method.

in Figure 4.8. Fluorolube has no bands at high wavenumbers, but many bands below 1300 cm^{-1} due to CF and CCl stretching and bending vibrations. Nujol has only one small band below 1300 cm^{-1}, so Nujol and Fluorolube mask different parts of the infrared spectrum. In the split mull technique, spectra are obtained of the sample dispersed in each mulling agent. The spectra are then spliced together, which can be done with computer programs or by literally cutting and pasting the spectra together. The spliced spectrum contains no bands from the mulling agents, only bands from the sample. Figure 4.9 shows the spectra of Nujol and Fluorolube spliced together. Note the absence of any strong infrared bands save those of carbon dioxide and water vapor (Nujol has a small band at 720 cm^{-1}). A problem with the split mull technique is that by the time the two mulls are prepared, and the spectra are obtained and spliced together, it may be faster to obtain the spectrum using a KBr pellet.

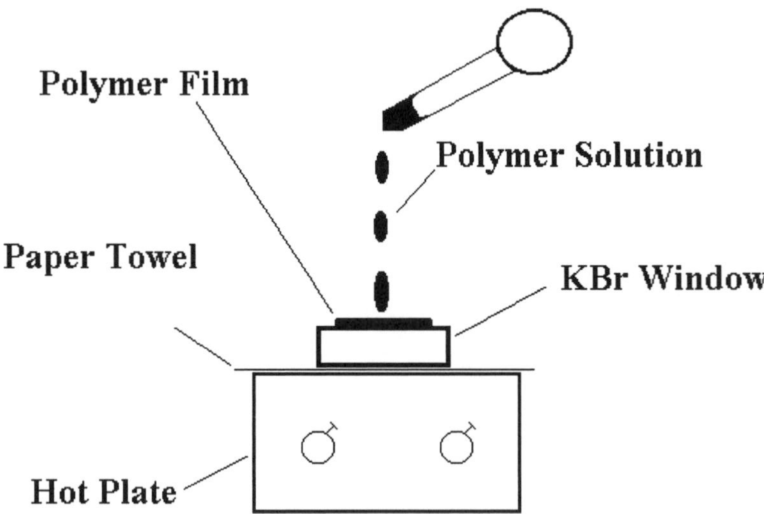

Figure 4.10 Diagram of how to cast a polymer film. The polymer solution is dripped onto the KBr window at intervals, while the hot plate gently heats the window to speed evaporation of the solvent.

Mulls are typically used for qualitative work, but not for quantitative work, since concentrations and pathlengths are difficult to reproduce using mulls.

Cast Films

Cast films are a way of preparing polymers for infrared analysis. Cast films are made by dissolving the polymer in an appropriate solvent, then placing drops of the solution on a KBr window and allowing the solvent to evaporate. A film of the polymer forms on the KBr window as the solvent evaporates. Drops of polymer solution are placed on the window at intervals to slowly build up a polymer film of the right thickness. The KBr window and the thin film are then placed in the infrared beam. The background should be run on the same clean KBr window that the polymer is deposited upon. A way of speeding up evaporation of the solvent is to place the KBr window and a paper towel on a hot plate set on low. Then the polymer solution is slowly dripped onto the KBr window. This is illustrated in Figure 4.10. Do not overheat the window, since it will crack due to thermal stress, and the paper towel may be set on fire (the author speaks from experience here!). Figure 4.11 shows the infrared spectrum of a cast film of an ethylene/vinyl acetate copolymer. The film was cast from a methylene chloride solution.

The key to making quality cast films is to find a solvent that will dissolve the polymeric sample and still have a low enough boiling point to evaporate

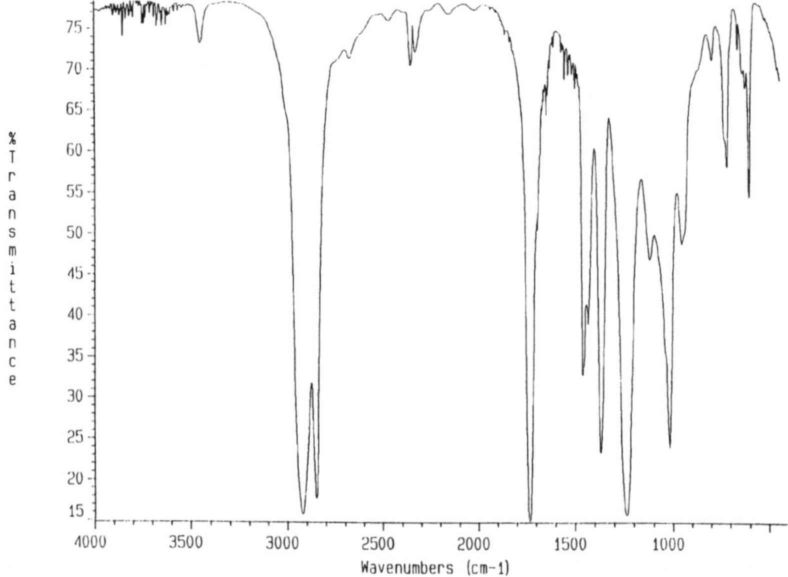

Figure 4.11 The infrared spectrum of an ethylene/vinyl acetate copolymer. The sample was prepared as a cast film deposited from methylene chloride solution.

easily. If the right solvent cannot be found, the polymer cannot be analyzed using this technique. The drawback of cast films is that they can be time consuming. Preparing the solution and waiting for the solvent to evaporate takes time. It is also not obvious when the polymer film is of the right thickness. Finally, by dissolving and precipitating the sample, the crystallinity and morphology of the polymer are changed. If this information is important to you, do not use the cast film technique. Quantitative work is rarely done using cast films, since their preparation can be very irreproducible.

Heat and Pressure Films

Another way of preparing polymer films is called the *heat and pressure* method. In this technique the sample is placed between the plates of a special hydraulic press that can simultaneously apply heat and pressure to a sample in a controlled fashion. The polymer is typically placed between sheets of aluminum foil, or better yet, teflon coated foil to prevent sticking. The goal is to heat the sample to a temperature above which it will flow, and the pressure will cause the sample to form a thin film. Usually 5 to 15 minutes of applying heat and pressure is sufficient to obtain the desired film thickness of around 20 microns. After removing the heat and pressure, the film is allowed to cool, and is analyzed by simply taping the film to the sample slide mount in an

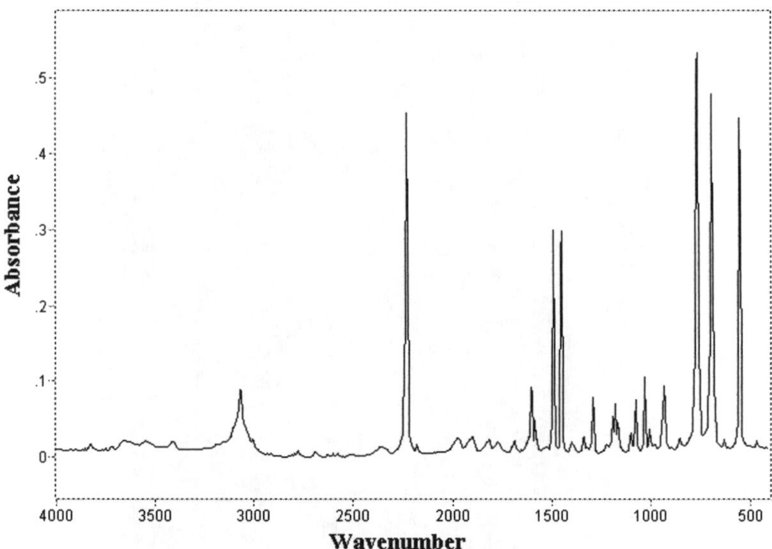

Figure 4.12 The infrared spectrum of benzonitrile obtained using the capillary thin film method.

FTIR sample compartment (make sure the infrared beam doesn't pass through the tape!).

The advantages of the heat and pressure method are that it will work on any sample of polymer that will flow at a temperature of less than 300° C. Also, irregularly shaped samples such as plastic pellets and plastic parts can be analyzed. Therefore, the heat and pressure method is a much more universal technique than the cast film method. A drawback to the heat and pressure method is that one does not know what film thickness will be obtained using a given temperature and pressure. This means some trial and error is involved in finding the right settings. Another disadvantage of applying heat and pressure to a polymer is that it changes the sample, effecting polymer crystallinity and morphology. If one is worried about the sampling technique changing anything about the sample, don't use this technique. Quantitative work is rarely done using heated and pressed films, since their preparation can be very irreproducible.

A.2 Transmission Spectra of Liquids

Capillary Thin Films

The simplest way to obtain transmission spectra of organic liquids is to place a drop or smear of the sample on a KBr window, place a second window on top of the first, and place the resulting "sandwich" directly in the infrared

Figure 4.13 A demountable infrared sampling cell. The two KBr windows pictured in front are held in place by the black screw on ring shown attached to the sample holder. (Photo courtesy of Spectra-Tech Inc.)

beam. This is known as the *capillary thin film* method. The name derives from the fact that the capillary action of the liquid holds the KBr windows together. The advantages of the capillary thin film technique is that it produces quality spectra of liquids in a fast and easy manner. It takes only a few seconds prepare samples using this technique. The background spectrum should be run with the same two KBr windows used in the thin film sandwich. Figure 4.12 shows a capillary thin film spectrum of benzonitrile.

There are some problems with the capillary thin film method. Volatile liquids will evaporate to some extent while their spectrum is being obtained. If the liquid being sampled is toxic or smelly, you may not want to risk exposing yourself or co-workers to the sample's vapors. Sample evaporation also makes capillary thin films unsuitable for quantitative analysis; it is difficult to measure concentration if the sample is disappearing in front of your very eyes. To avoid sample evaporation, a sealed liquid cell (see below) should be used.

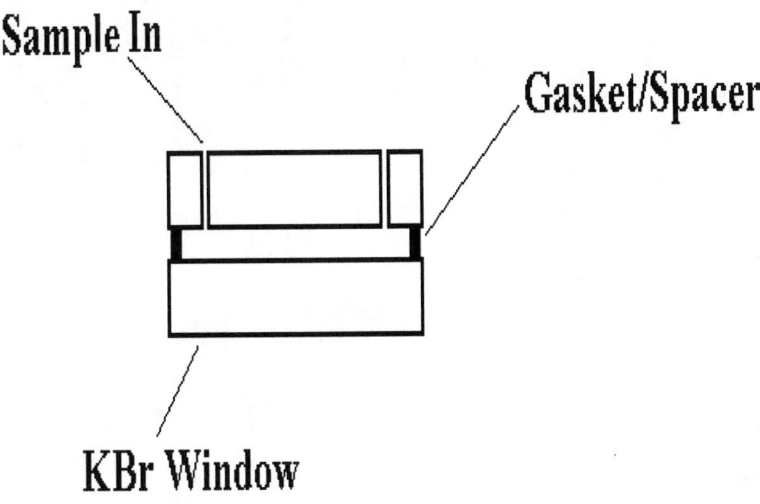

Figure 4.14 A schematic diagram of a sealed liquid cell.

Lastly, the capillary thin film technique is difficult to perform on liquids containing water. Water will dissolve the KBr windows, and water bands are difficult to subtract out of sample spectra. Window materials such as ZnSe or AgCl do not dissolve in water, and can be used for capillary thin film spectra of aqueous samples. However, a technique described below called liquid ATR is much better for analysis of aqueous solutions.

An excellent method of obtaining capillary thin film spectra is to use what is called a *demountable cell holder*. A photo of a demountable cell is shown in Figure 4.13. It consists of a metal plate that slides into the FTIR sample mount, with a hole over which two infrared transparent windows can be placed. The sample is placed between the windows, and a screw ring is used to apply pressure to hold the sample and windows together . The advantage of applying pressure is that thick samples are brought to the proper thickness, the evaporation of liquid films can be slowed, and semisolid samples can be pressed and turned into films. Finally, the demountable cell holder does a good job of reproducing the position of the sample in the infrared beam.

Sealed Liquid Cells

Sealed liquid cells are another way of sampling liquids in transmission. A

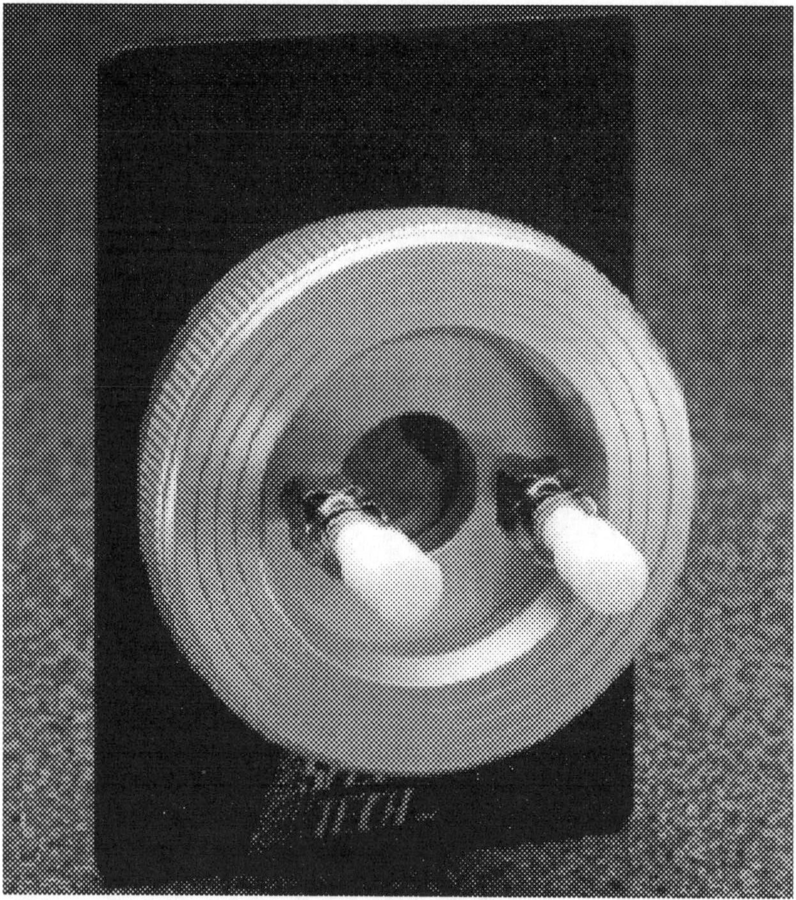

Figure 4.15 A photo of a sealed liquid cell. The two white plugs prevent the sample from evaporating. (Photo courtesy of Spectra-Tech Inc.)

schematic diagram of a sealed liquid cell is shown in Figure 4.14, and a picture of an assembled cell is seen in Figure 4.15. The cells consist of two KBr windows, one of which has two small holes drilled into it. The two windows are separated by a gasket of a specific thickness. The gasket material can be teflon, or a lead/mercury amalgam. The thickness of the gasket determines the pathlength of the cell and also prevents the sample from evaporating. The sample is placed into the cell through holes drilled in the windows. Since the holes are small, the cell is usually filled using a syringe. Cells with thicknesses from 1 to 100 microns are typical. The entire cell assembly is held in a metal frame by screws and a retaining plate, and the metal frame slides into the sample mount in the FTIR. Plugs keep the sample from evaporating through

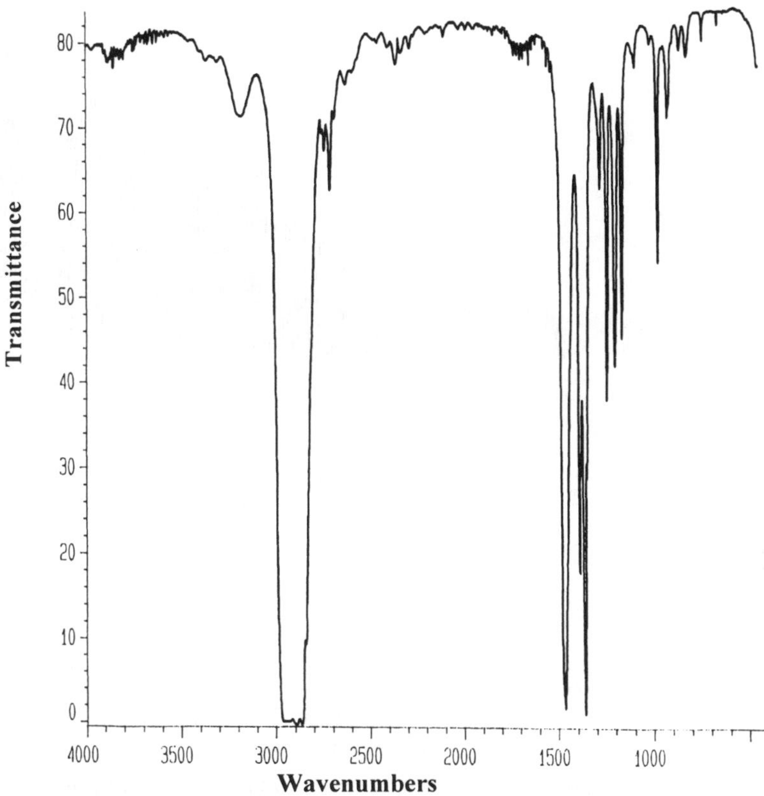

Figure 4.16 The infrared spectrum of isooctane obtained using a sealed liquid cell with a 25 micrometer pathlength.

the holes in the window. The background spectrum is run of the same clean, empty liquid cell as will be used in the analysis of the sample. The spectrum of isooctane obtained with a sealed liquid cell is shown in Figure 4.16. If KBr windows are used in sealed liquid cells, the samples must not contain any water since KBr is water soluble. If you need to analyze aqueous samples using this method, cells made from ZnSe or AgCl should be used since these materials are not water soluble. However, water is a strong infrared absorber and will tend to hide much of the spectrum of any material dissolved in it.

The advantage of sealed liquid cells is that they are excellent for quantitative work, since the pathlength is known. They also prevent the sample from evaporating, so volatile or toxic liquids can be sampled. The major disadvantage of sealed liquid cells is that they are difficult to fill and clean. The tiny holes drilled in the KBr windows prevent the sample or cleaning solvents from being conveniently introduced into the cell. It is often times necessary

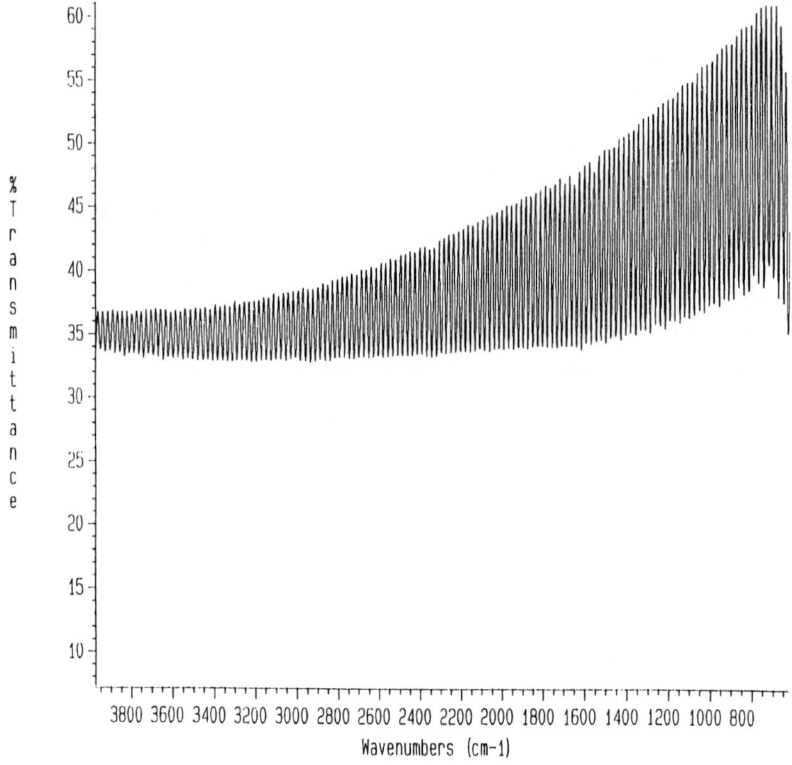

Figure 4.17 A spectrum of an empty sealed liquid cell showing interference fringes. The distance between adjacent fringes can be used to calculate the pathlength of the cell.

to flush a sealed liquid cell with solvent a number of times to remove all the sample. Also, it is easy to overfill the cell and spill sample on the salt windows. Before using a sealed liquid cell, it is a good idea to obtain the spectrum of the cell using it as the sample and using air as the background. Any contaminants in the cell will give bands in this spectrum, and should be removed by cleaning. For qualitative work on liquids, the capillary thin film method discussed above is generally easier and faster than using sealed liquid cells. However, sealed liquid cells are a necessity for quantitative analysis of liquids.

In addition to running the spectra of liquids, the spectra of solids can be obtained using sealed liquid cells. The solid is dissolved in an appropriate solvent, and the solution is placed in the sealed liquid cell. The choice of solvent is critical. It should dissolve the sample, but not react with it. Unfortunately, all solvents have bands in the infrared which can mask the spectrum of the sample. A large part of solvent choice is finding a solvent that dissolves the sample without masking the sample's infrared bands. Common IR solvents are carbon tetrachloride and carbon disulfide. These two solvents mask totally different parts of the infrared spectrum, and by obtaining the spectrum of a sample in both solvents and splicing the spectra together, a spectrum mostly free of solvent bands can be obtained. A listing of infrared transparent solvents and their useful wavenumber ranges is found in ref. [1].

When the spectrum of an empty liquid cell is obtained it gives rise to sinusoidal variations in the baseline known as *interference fringes.* An example of interference fringes is shown in Figure 4.17. Interference fringes are caused by interference between light rays that pass through the liquid cell, and rays that reflect off the inside surfaces of the cell. Interference fringes appear in the infrared spectrum as a cosine wave. Fringes are useful since they can be used to accurately calculate the pathlength of a sealed liquid cell. To do this, measure the difference in wavenumber between two maxima in a fringe pattern, then use the following formula:

$$D = N/2(W_1 - W_2) \qquad (4.1)$$

where:

D = Cell Thickness
N = Number of maxima between two wavenumbers
W_1 = Wavenumber 1
W_2 = Wavenumber 2

The problem with interference fringes is that they may appear in sample spectra and obscure important bands. Often times introducing the sample into the cell will suppress the interference fringes. However, the best way to avoid fringes is to perform the background scan on an empty, clean liquid cell. The fringes should ratio out when the sample spectrum is obtained.

A3. Transmission Spectra of Gases

A diagram of the type of cell most commonly used to obtain the spectra of gases is shown in Figure 4.18. The cell is typically 10 cm long, consists of a glass or metal body, and has KBr windows on both ends to let the infrared beam pass through. The cell has valves to allow gases in and out, and the windows are equipped with gaskets to prevent gas leakage. Gases are introduced to the cell using a vacuum manifold consisting of a vacuum pump and appropriate plumbing. The sample is contained in a lecture bottle, glass bulb, or gas cylinder connected to the manifold. To introduce a gas into the cell, the manifold and cell are first evacuated. An in-line pressure sensor should be used to measure the pressure in the cell. If one cannot achieve a low pressure in the cell, it indicates a leak is present. Try tightening the rings holding the gaskets and windows, or examining the valves or connection to the vacuum pump to see if they are leaking. After evacuating the gas cell, the pump is isolated from the manifold by closing valves, and the vessel containing the gas of interest is opened. Gas is allowed to leak into the manifold and cell

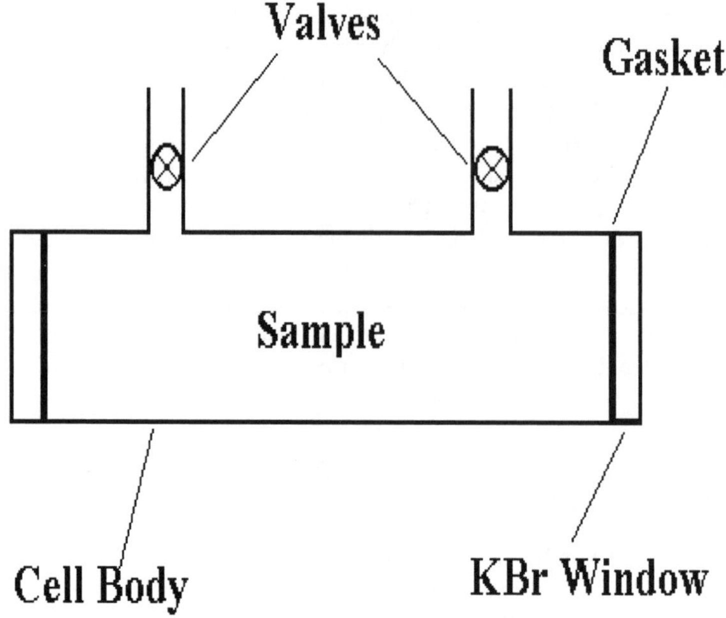

Figure 4.18 A diagram of a 10 cm gas cell.

until the desired pressure is reached. The valve on the gas vessel is then closed. The final pressure of gas in the manifold and cell should be noted. The cell valve is closed, and the valve to the pump is opened to allow it to evacuate the manifold. The gas cell can now be placed on a special holder that slides into the sample slide mount in the instrument's sample compartment. The background spectrum should be run on an evacuated gas cell. If trace amounts of gas need to be detected, gas cells are available with pathlengths ranging from 1 meter to several kilometers. These cells use ingenious optical designs to obtain long pathlengths but still keep the cells a manageable size. Concentrations as low as parts per billion can be detected using these gas cells.

A technique known as *grab sampling* can also be performed with gas cells. An evacuated gas cell is taken into the field to a site where samples are to be collected. The valve on the cell is opened, and some of the local atmosphere is sucked into the cell. The cell is then taken back to the lab for FTIR analysis. Lastly, *flow through* analysis is also possible with gas cells. Both valves on the cell are left open, and the sample gas is allowed to continuously flow through the cell. This technique can be used to look at gases given off by

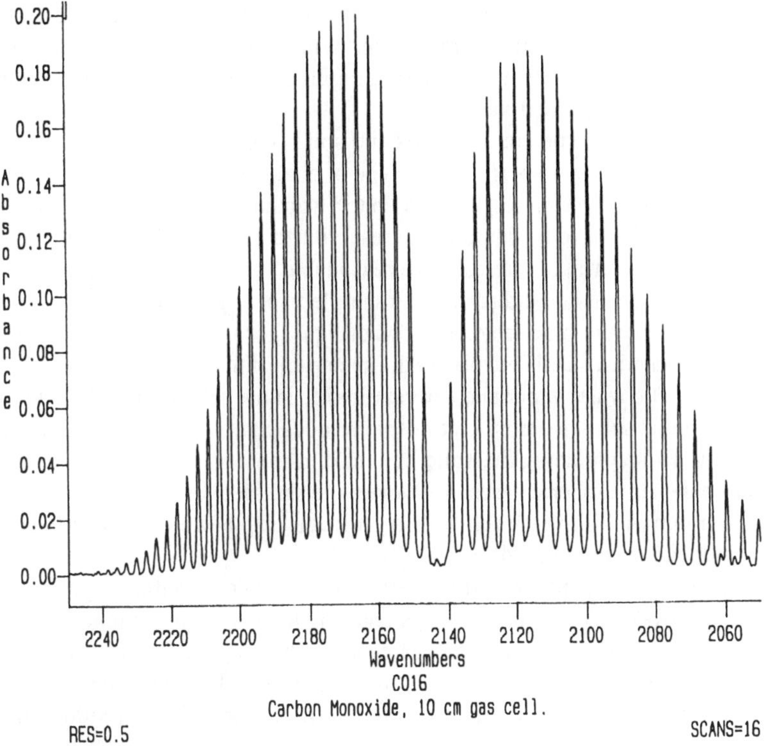

Figure 4.19 The infrared spectrum of carbon monoxide (CO). Conditions were 16 torr of gas and 10 cm pathlength.

chemical reactions, and has also been used to monitor the gaseous contents of smoke stacks.

Gas phase spectra look totally different than the spectra of solids and liquids. Gas phase molecules are separated by large distances (at least on a molecular scale). As a result, there is very little interaction between gaseous molecules. It is the molecular interactions between closely spaced molecules in liquids and solids that cause their infrared bands to be relatively wide (typically > 10 cm^{-1} wide). Since molecular interactions in gases are weak, their infrared bands are very sharp. This is shown in Figure 4.19, which shows the infrared spectrum of carbon monoxide. The many lines seen in Figure 4.19 all comprise one infrared band. The lines are due to transitions from rotational levels in the vibrational ground state to rotational levels in the excited vibrational state. These transitions are known as *rovibrational transitions*, and are the predominant feature in gas phase spectra. As a result, their are many sharp lines in gas phase infrared spectra. If gas and liquid phase spectra are obtained of the same molecule, the narrow gas phase lines would be seen to grow together and collapse into one broad band in the liquid spectrum. However, the center of the infrared band will be about the same in the two spectra. This is precisely what is seen in Figure 2.16, which shows the spectra of liquid water and water vapor.

The vapor phase spectra of liquids and solids can be obtained using a gas phase cell. One way to perform this type of analysis is to have a reservoir of the solid or liquid of interest hooked up to a vacuum manifold, and allow the vapor above the solid or liquid in the reservoir to pass into the gas cell. This assumes the material of interest has a reasonable vapor pressure at room temperature. The quick and dirty way to sample vapors given off by solids and liquids is to simply place several drops of liquid, or several mg of solid, into an unevacuated gas cell, and obtain the spectrum of the vapor above the solid or liquid. The water vapor and carbon dioxide in the cell will interfere, but if the sample absorbs where the atmosphere does not, it won't be a problem. One can also try placing the solid or liquid in the cell, and evacuating it to remove the atmospheric gases. However, the sample may be sucked into the vacuum pump! Filling the cell with dry nitrogen can be used to purge the cell of water and carbon dioxide.

FTIR is an excellent method for determining the composition of unknown gases. FTIRs today are being used extensively in the field in conjunction with gas cells to monitor air quality and catch air polluters. The future may see the development of an FTIR air monitor that makes use of a small, rugged FTIR in tandem with a small, rugged gas cell to monitor the quality of indoor air.

B. Reflectance Techniques

Reflectance sampling techniques differ from transmission techniques in that the infrared beam is bounced off the sample instead of passing through the sample. When discussing reflectance, it is convenient to define a line called the *surface normal*, which is drawn perpendicular to the surface of the sample. An example of a surface normal is shown in Figure 4.20. The angle of incidence of a light beam is defined as the angle the incoming light ray makes with the surface normal. The angle of reflectance is defined as the angle the outgoing light beam makes with the surface normal. There are two different types of reflection as illustrated in Figure 4.20. *Specular reflectance* occurs when the angle of incidence equals the angle of reflectance. This is the type of reflection that takes place off of smooth surfaces, such as that of mirrors. *Diffuse reflectance* occurs when the angle of incidence is fixed, but angles of reflection vary from 0 to 360 degrees. Diffuse reflectance occurs on rough surfaces, which are the most common surfaces found. Both types of reflectance can be used to obtain the infrared spectra samples, as will be seen below.

There are several disadvantages of reflectance sampling techniques compared to transmission techniques. First of all, reflectance techniques require special accessories that must be placed into the spectrometer's sample compartment. These accessories will either mount on the baseplate in the sample

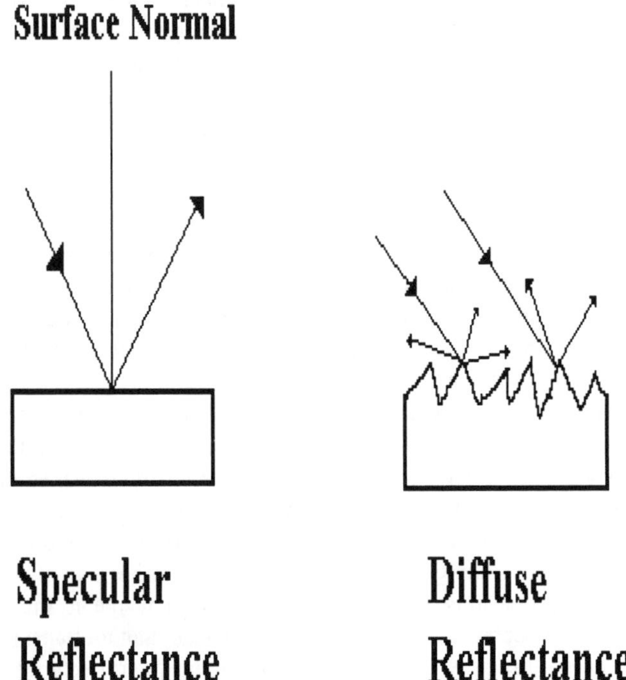

Figure 4.20 Examples of specular and diffuse reflectance. In specular reflectance the angle of incidence equals the angle of reflection. In diffuse reflectance, the reflected light leaves the sample in any and all directions.

compartment, or in the sample slide mount. The accessories contain mirrors to focus the infrared beam to the sample and to collect the reflected light and send it to the detector. Reflectance accessories cost $1000 or more, which is more expensive than the tools used for transmittance sample preparation. The second disadvantage of reflectance methods is that the depth to which the infrared beam penetrates into a sample is not accurately known. In a reflection experiment the depth of penetration of the infrared beam is determined by the properties of the sample, the properties of the surface, and the angle of incidence of the light. As such, it is difficult to exactly determine the pathlength in a reflectance experiment, making quantitation difficult since pathlength is an important variable in determining a sample's absorbance (see Chapter 5).

In a reflection experiment, typical depths of penetration are 1 to 10 microns. Since this is such a small distance, the sample's surface contributes most strongly to the spectrum. This is in contrast to transmission techniques, where the bulk of a sample rather than the surface contributes most strongly to the spectrum. The fact that reflectance techniques obtain the spectrum of the surface of a sample is good if the surface is of interest. However, the chemical composition of a sample's surface can be significantly different from that of the bulk, so the transmission and reflection spectra of the same sample can look totally different.

It is difficult to capture all the light reflected off of a sample's surface, so reflection spectra may be noisier than transmission spectra for a given number of scans and resolution. As a result, it may be necessary to use more scans with reflectance samples to get the SNR to an appropriate level. Lastly, reflectance spectra of solids and liquids can be obtained, but there is no way of obtaining reflectance spectra of gases. Thus, reflectance techniques are not as universal as transmission techniques.

After the lengthy discussion of the disadvantages of reflectance techniques, the reader may be left wondering why anyone uses these techniques. One advantage of reflectance techniques over transmittance methods is there is no thickness problem. One need not worry about whether the sample is the proper thickness or concentration, or whether the proper amount of light is passing through the sample. Therefore, one does not have to spend a lot of time diluting, squashing, or grinding the sample. This means the sample preparation for reflectance samples is faster and easier than for transmittance samples. Often, the time saved in sample preparation using reflectance techniques is of greater dollar value than the price of a reflectance accessory. A final advantage of some reflectance techniques is that they are nondestructive. The sample is left intact after its spectrum is obtained, which means the sample can be used for other analyses.

The time advantage and ease of use of reflectance techniques is making this type of analysis more and more popular. The author's personal bias is to use reflectance techniques as a first choice for most samples to save on time. If the reflectance technique does not produce a good spectrum, then transmission techniques are used.

The following discussion will focus on diffuse reflectance (DRIFTS) which is used on solids, attenuated total reflectance (ATR) which is used on films, semisolids, and liquids, and specular reflectance which is used on smooth surfaces.

Diffuse Reflectance (DRIFTS)

Diffuse Reflectance Infrared Fourier Transform Spectroscopy (DRIFTS) is used to obtain the infrared spectra of powders and other solid materials. It is a technique that was made possible with the advent of FTIR [2], and as such does not have as much history behind it as transmission techniques such as KBr pellets and mulls. However, it does have some advantages over transmission techniques, as will be discussed below.

The sample preparation for DRIFTS is similar to that for KBr pellets. The sample and KBr should be ground separately to reduce particle size, either in

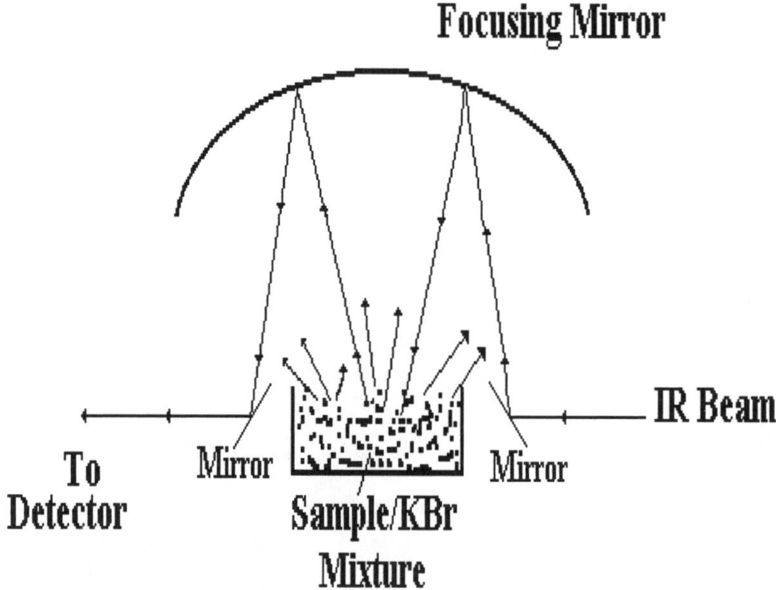

Figure 4.21 The optical diagram of a diffuse reflectance (DRIFTS) accessory.

an agate mortar and pestle or Wig-L-Bug (see above). The ground sample is diluted to 1 to 10% in the ground KBr. A spatula should be used to stir the mixture to ensure the sample is properly diluted. The spatula is then used to spoon the mixture into a sample cup. The cups are usually made of metal, and are generally 1/2" in diameter and 1/8" deep. The filling of the sample cup is actually very important. The intensity of diffusely reflected light is dependent upon the packing density of the particles in the sample. So, filling the sample cup in different ways can lead to different packing densities and different band heights. The author suggests you fill the sample cup to overflowing using the spatula, and then use the edge of the spatula like a knife blade and drag it across the top of the sample cup to remove the excess powder. When you are done, the surface of the powder should be level, and the sample cup should be full up to the brim. Do not tap the sample cup against anything to make the particles settle. The intensity of diffusely reflected light also depends on the particle size of the sample. Tapping causes the big particles to rise to the top, presenting an atypical sample surface to the infrared beam.

Next, the sample cup is placed at the focal point of a diffuse reflectance accessory. There is no other sample preparation involved. An optical diagram of a DRIFTS accessory is shown in Figure 4.21. DRIFTS accessories

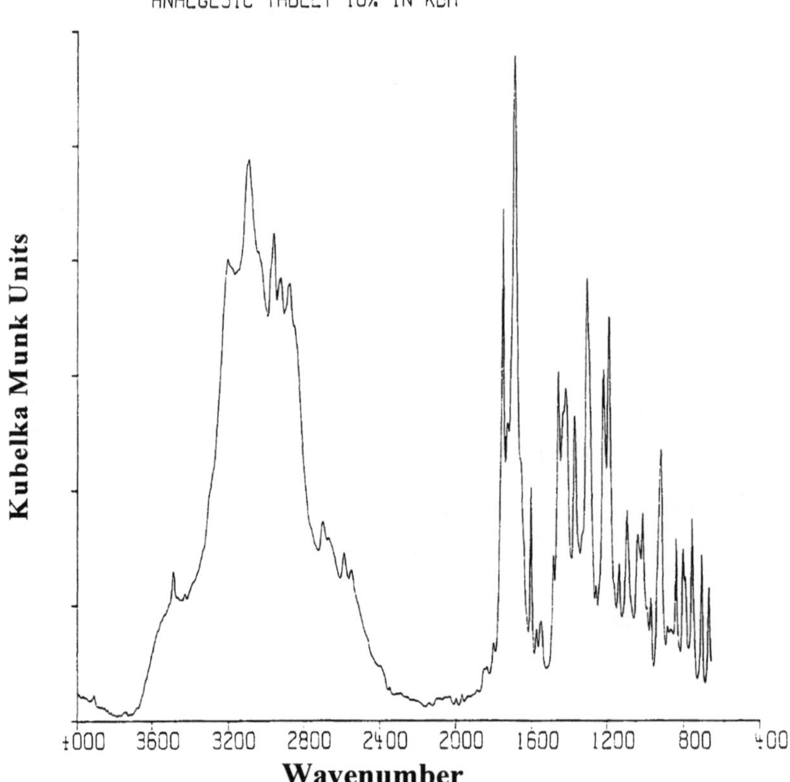

Figure 4.22 The diffuse reflectance (DRIFTS) spectrum of an analgesic tablet. The sample was diluted 10% in KBr. Note that the Y axis is in Kubelka-Munk units.

either drop into the sample compartment baseplate, or are placed into the sample slide mount. The accessory consists of flat mirrors that direct the incoming radiation onto a spherical or ellipsoidal focusing mirror. The radiation is focused on the powdered mixture in the sample cup, and is diffusely reflected into a 360° circle. Diffusely reflected radiation is made up of light that is scattered, absorbed, transmitted, and reflected by the sample. As much of the diffusely reflected light as possible is collected since it has interacted with the sample, and carries information about the sample. A second spherical or ellipsoidal mirror collects the diffusely reflected radiation. The light reflects off more flat mirrors, and is focused on the detector. The background spectrum is obtained from pure, ground KBr placed in a DRIFTS sample cup. An example of a DRIFTS spectrum is seen in Figure 4.22, which is the spec-

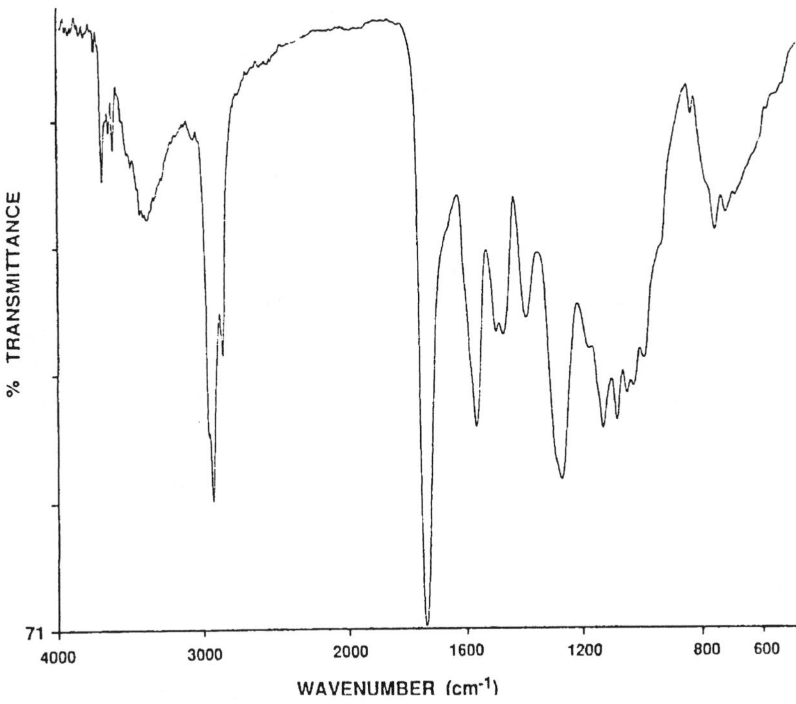

Figure 4.23 The DRIFTS spectrum of paint on an aerosol can obtained by scraping the paint off with silicon carbide paper, then placing the paper in a DRIFTS accessory.

trum of an analgesic tablet diluted 10% in ground KBr. High quality DRIFTS spectra are normally obtained using from 10 to 100 scans.

The real advantage of DRIFTS is that it does not require the pressing of a pellet. Therefore, sample preparation is much faster and simpler than with KBr pellets. Also, DRIFTS can be used on a wider variety of samples than KBr pellets. The major disadvantage of DRIFTS is the cost of the accessory, which can be more than $1000. Diffuse reflectance accessories can be purchased from some instrument manufacturers, or from companies that specialize in the manufacture of FTIR accessories. Best results will be obtained with an accessory that is designed for use in your FTIR. If the optics of the accessory and the FTIR are not matched, you will be throwing away valuable light energy because the infrared beam will not be focused properly on the sample. A listing of FTIR accessory manufacturers is given at the end of this chapter.

The applications of DRIFTS are quite widespread in the pharmaceutical and chemical industries. In addition to sampling milligram quantities of sample, smaller sample amounts down to a microgram can be detected using DRIFTS. This "micro DRIFTS" technique is used in one of two ways. The sample can be ground then sprinkled on top of a sample cup already filled with ground KBr, concentrating the sample where the IR beam is most likely to interact with it. If the amount of material for analysis is extremely small, microscale sample cups are available. The sample is simply placed in the cup with no grinding or diluting in KBr, and the sample can be removed afterwards and used in other analyses.

Very large and intractable objects can also be analyzed by DRIFTS. Examples include pieces of furniture, large pieces of plastic, and table tops. The difficulty in analyzing such samples is that they won't fit into the sample compartment, and it may be difficult or undesirable to grind or melt the sample. The solution to the problem is to take a piece of silicon carbide paper and rub it on the sample and accumulate some particles. The paper containing the sample particles is cut out and placed in the DRIFTS cup, then the spectrum is obtained. Some accessory companies sell silicon carbide disks that are the same size as the DRIFTS sample cup. This disk is rubbed on the sample then placed directly in the sample cup, saving some time. The background spectrum is obtained on the clean silicon carbide paper. Figure 4.23 shows the spectrum of paint coated on the outside of an aerosol can, which was obtained in this manner.

DRIFTS can also be used to sample liquids. Drops of a solution can be dripped into a sample cup filled with ground KBr. If the solvent is volatile it will evaporate, leaving the solute behind. As more and more drops are evaporated, the solute concentrates in the KBr, so small amounts of sample can be concentrated and detected in this way. DRIFTS has been used in this manner to sample HPLC peaks. Drops of HPLC eluant are allowed to drip into a KBr filled sample cup, the sample is allowed to evaporate, and the spectrum of the HPLC peak can be obtained. By keeping an eye on the chromatographic detector, different sample cups can be used to collect different HPLC fractions.

Recall that reflectance techniques in general sample the surface rather than the bulk of a sample. DRIFTS is useful for studying the surface of powdered materials that consist of coated particles. Chambers for evacuating and heating samples are also available for DRIFTS accessories. These allow the controlled introduction of adsorbates onto powdered surfaces. This has enabled the surfaces of catalysts and adsorbents to be monitored under real world conditions.

Quantitative DRIFTS

DRIFTS is an excellent technique for qualitative analysis. However, quantitative analysis using DRIFTS requires strict attention to experimental detail. The equation used to relate concentration to peak heights (or areas) in diffuse reflectance spectra is called the Kubelka-Munk equation [3]. These spectra are plotted in Kubelka-Munk units and are referred to as Kubelka-Munk spectra. An example of a DRIFTS spectrum plotted in Kubelka-Munk units is shown in Figure 4.24. When making quantitative measurements using DRIFTS, the spectra <u>must</u> be in Kubelka-Munk units, just like the spectrum must be in absorbance units when performing quantitative measurements on transmission samples. The form of the Kubelka-Munk equation is as follows:

$$KM = (1 - R_{\infty})^2/2R_{\infty} \quad (4.2)$$

$$KM = k/s$$

where:

KM = Spectrum in Kubelka Munk Units
R_{∞} = Reflectance from an infinitely thick sample
k = Absorption coefficient
s = Scattering factor

The absorption coefficient k is defined as follows:

$$k=2.303ac \quad (4.3)$$

where:

a = Absorptivity
c = Concentration

Substituting equation 4.3 into equation 4.2 yields

$$KM = 2.303ac/s \quad (4.4)$$

In practice, the spectrum of the sample is recorded in percent transmission and is used as R_{∞}. The instrument's software then uses equation 4.2 to calculate the Kubelka-Munk spectrum. The Kubelka-Munk equation assumes the sample is infinitely thick compared to the depth of penetration of the beam into the sample. This is generally true for samples contained in cups that are 1/8" deep since the light only penetrates 1 to 10 microns into the sample. Note from equation 4.4 that the absorption coefficient (k) is directly proportional to the absorptivity (a) and concentration (c) of the sample, so the peak heights in a Kubelka-Munk spectrum vary linearly with concentration.

The real unknown quantity in equation 4.4 is the scattering factor, s. This quantity is dependent upon the particle size distribution, particle shape distribution, and packing density of the sample. These variables are very difficult to control, and are the main reason why quantitative analysis using DRIFTS can be difficult. For example, the way in which the sample is ground affects particle size and shape. The way the sample is packed into the DRIFTS sample cup affects packing density. Even tapping the sample cup once it has been filled can change the packing density enough to affect peak heights. Fortunately, with good technique there are ways of keeping the variability in the

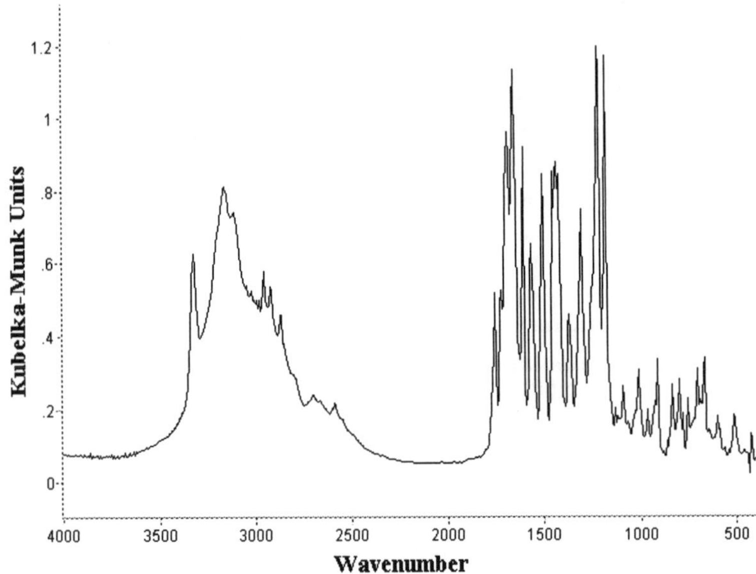

Figure 4.24 A DRIFTS spectrum plotted in Kubelka-Munk units.

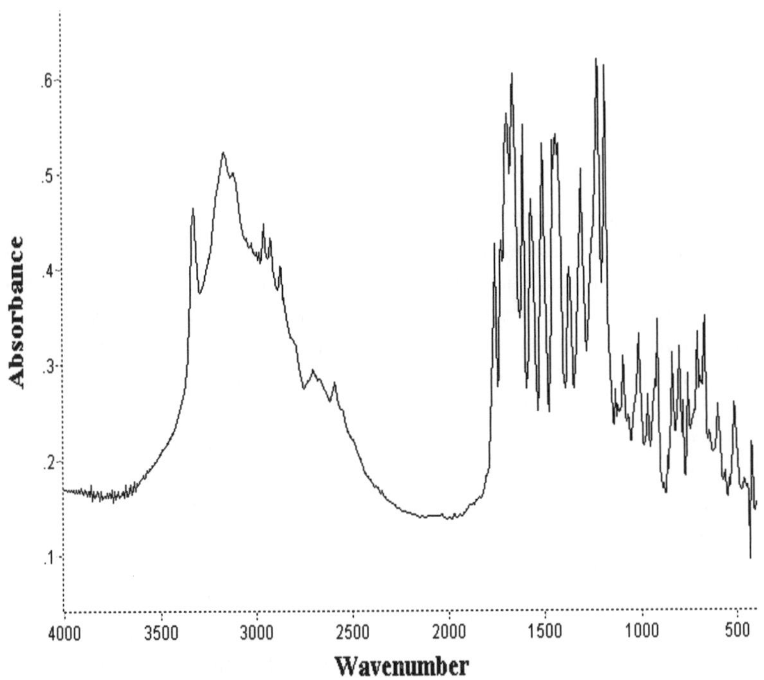

Figure 4.25 The same DRIFTS spectrum as in Figure 4.24 plotted in "diffuse absorbance" format. Spectra plotted in diffuse absorbance can be used for interpretation, but should not be used for quantitiative analysis.

scattering factor to a minimum so that quantitative analysis can be performed. Reproducible grinding is best accomplished through the use of a Wig-L-Bug (see above). Measure the amount of KBr or sample to be put into the grinding capsule, and always use the same amount for both (e.g., 100 mg). Be sure to grind the sample and KBr, and grind them for the same amount of time (one minute is usually sufficient). Measure the ground KBr into a second capsule (e.g., 100 mg), then measure the ground sample into this capsule (e.g., 5 mg). Then shake this capsule in the Wig-L-Bug, without using a ball bearing, for about a minute. In this way, the sample and KBr will be intimately mixed while no more grinding occurs. Spoon this mixture into the sample cup, then drag the edge of the spatula across the top of the sample cup to level off the sample. You are now ready to obtain the DRIFTS spectrum of the sample. The background should be run on the same ground KBr that was mixed with the sample. Measure all peak heights and areas from the Kubleka-Munk spectrum. After this, plot the calibration curve and treat the problem like any other quantitative analysis. As with many other sampling techniques, quantitative analyses using DRIFTS will be more accurate if peak area ratios or peak height ratios are used, and if the method of internal standards is used (see Chapter 5). By paying close attention to the reproducibility of the scattering factor, quantitative analyses with DRIFTS can be successful.

Although it is essential to use KM units for quantitative analysis in DRIFTS, other units can be used for displays of DRIFTS spectra if qualitative information is all that is desired. For instance, a "diffuse absorbance" spectrum, which is a DRIFTS spectrum plotted with absorbance as the Y axis, is a convenient way of viewing DRIFTS spectra for interpretation purposes. An example of a diffuse absorbance spectrum is shown in Figure 4.25. The only difference between Kubelka-Munk and diffuse absorbance spectra are the values on the Y axis. The peak positions (X axis) in diffuse absorbance and Kubelka-Munk spectra are identical. However, remember that diffuse absorbance spectra should not be used for quantitative analysis.

Attenuated Total Reflectance (ATR)

The Attenuated Total Reflectance (ATR) technique is used to obtain the spectra of solids, liquids, semisolids, and thin films. ATR is performed using an accessory that mounts in the sample compartment of an FTIR. A schematic diagram of such an accessory is shown in Figure 4.26. At the heart of the accessory is a crystal of infrared transparent material of high refractive index. Typical materials used are zinc selenide, KRS-5 (thallium iodide/thallium bromide), and germanium. Mirrors on the accessory bring the infrared radiation to a focus on the face of the crystal. After passing through the crystal and reaching its top surface, you would expect the radiation to leave the crystal. However, if the crystal has the proper refractive index and the light has the proper angle of incidence, the radiation undergoes what is called *total internal reflection.* The infrared energy reflects off the crystal surface rather than leaving the crystal. In Figure 4.26 the infrared beam reflects off the

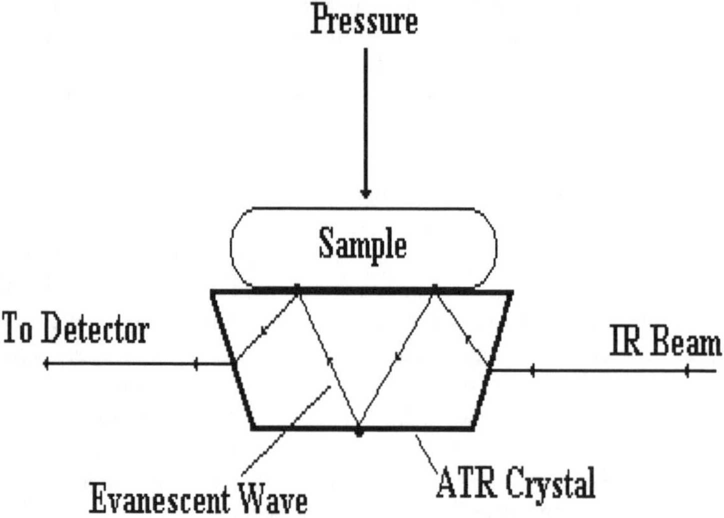

Figure 4.26 A schematic diagram of an attenuated total reflectance accessory.

crystal surface three times before leaving the crystal. In this instance the crystal acts as a waveguide for the infrared radiation, which will follow the shape of the crystal the same way water follows the shape of a hose in which it is flowing. Total internal reflection is the phenomenon that allows light to travel through fiber optic cables, and makes fiber optic communications possible.

Once the infrared radiation is inside the crystal, a standing wave of radiation is set up, called an *evanescent wave.* A unique property of the evanescent wave is that it is slightly bigger than the crystal, and so it penetrates a small distance beyond the crystal surface into space. This is illustrated in Figure 4.26, where a small amount of the IR beam is shown penetrating above and below the crystal. A sample brought into contact with the crystal can interact with the evanescent wave, absorb infrared radiation, and have its infrared spectrum detected. The evanescent wave is attenuated by the sample's absorbance, hence the name *attenuated total reflectance* (ATR). Good contact between the sample and the crystal is critical to ensure the evanescent wave penetrates into the sample. This is why it is imperative that the crystal be kept clean and scratch free. Also, pressure is sometimes applied to samples to flatten them against the crystal to achieve good sample/evanescent wave coupling.

How deeply the infrared radiation will penetrate into a sample is known as the *depth of penetration* (DP), and is analogous to the idea of pathlength in

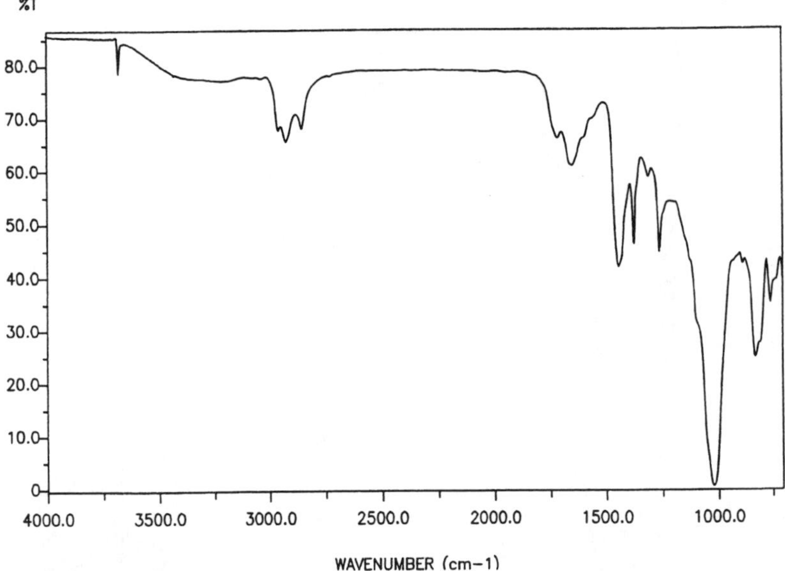

Figure 4.27 The ATR spectrum of a latex glove. Note that the bands at low wavenumber are much more intense than the bands at high wavenumber.

transmission sampling techniques. Depth of penetration is defined as the depth at which the evanescent wave is attenuated to 36.8% (1/e) of its total intensity. DP is given by the following equation:

$$DP = \frac{1}{2\pi W N_c (\sin^2\Theta - N_{sc}^2)^{1/2}} \quad (4.5)$$

where:

DP = Depth of penetration
W = Wavenumber
N_c = Crystal refractive index
Θ = Angle of incidence
$N_{sc} = N_{sample}/N_{crystal}$

Penetration depths on the order of 0.1 to 5 microns are typical. Equation 4.5 is complicated, but there are a few simple things about it that should be remembered. First, the depth of penetration is dependent on wavenumber. The DP goes down as wavenumber goes up. Thus, low wavenumber light penetrates further into the sample than high wavenumber light. As a result,

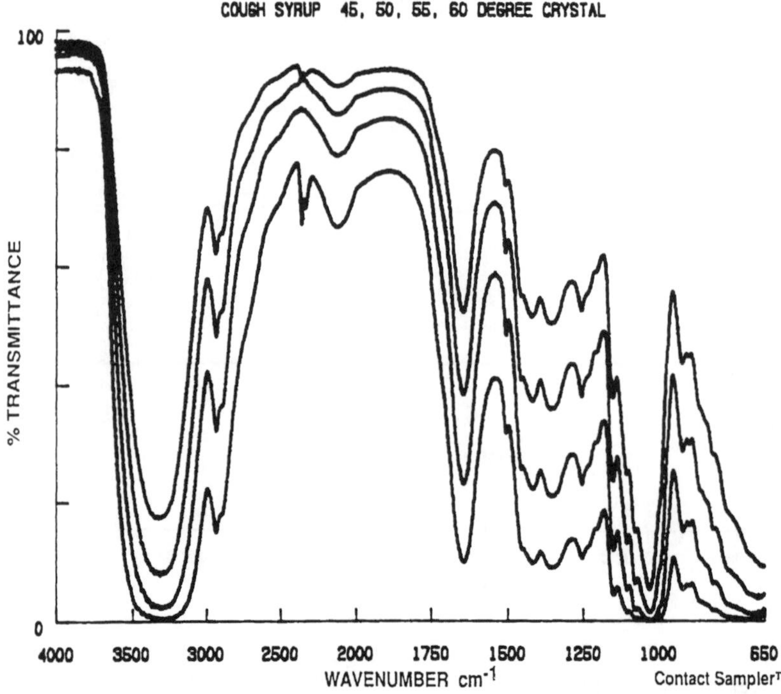

Figure 4.28 ATR spectra of a sample of cough syrup. The intensity of the bands decreases going from the botton spectrum to the top spectrum because the bevel angle of the crystal was changed. This changes the angle of incidence of the radiation, which in turn determines the depth to which the light penetrates into the sample.

ATR spectra show peaks that are more intense at low wavenumber than at high wavenumber. An example of this phenomenon is seen in Figure 4.27, which is the ATR spectrum of a latex glove. Note that the bands at low wavenumber are more intense than the bands at high wavenumber. The dependence of DP on wavenumber causes the relative peak intensities for ATR spectra and transmission spectra of the same sample to be different. Because of this, the best library searching results are obtained when ATR spectra are searched against libraries of ATR spectra rather than libraries of transmission spectra. Several instrument manufacturers sell software that perform a calculation on ATR spectra to remove the depth of penetration dependence.

The second thing to note about equation 4.5 is that the depth of penetration goes down as the refractive index of the ATR crystal goes up. Thus, Ge with a refractive index of 4.0 has a significantly shallower depth of penetration than a ZnSe crystal with a refractive index of 2.5. Changing crystal materials allows one to obtain spectra from different depths in a sample, which is known as *depth profiling.* If crystals of different refractive index are used to obtain spectra of the same sample, then the change in composition with depth in the sample can be obtained. This can be useful for looking at laminated films which consist of layers of different composition, or any sample where the change in composition with depth is of interest.

The denominator in equation 4.5 also includes the angle of incidence. Depth of penetration decreases as the angle of incidence of the infrared beam increases. The angle of incidence can be altered by changing the angle of the bevel of the ATR crystal, or by varying the angle of the incoming radiation. Some ATR accessories have movable mirrors that change the angle of incidence so DP can be varied controllably. An example of how changing the angle of incidence effects depth of penetration is shown in Figure 4.28. This figure shows the ATR spectrum of cough syrup taken using crystals with different bevel angles. Note how the intensity of the bands changes from the bottom spectrum to the top spectrum as the bevel angle of the crystal is changed.

Lastly, it should be noted that equation 4.5 is dependent on the refractive index of the sample. Fortunately, the refractive index of most organic compounds is very similar, so that DP varies little from sample to sample, all other factors being equal. Therefore, to a first approximation the depth of penetration in ATR can be considered to be sample independent. This approximation is very useful when performing quantitative analysis on samples of the same or similar refractive index, since the depth of penetration (i.e., pathlength) will be controlled as long as the angle of incidence and crystal refractive index are held constant. Since pathlength is one of the important variables in quantitative analysis, it is nice to have this variable controlled within the design of the sampling apparatus. This makes ATR a widely used technique for quantitative analysis.

Despite the complexity of equation 4.5 and the ensuing theoretical discussion, ATR is an easy to use sampling technique. ATR accessories mount in the sample compartment or slide mount of an FTIR. The background spectrum is obtained of the accessory containing the clean, dry crystal. The sample is brought into contact with the crystal, then its spectrum is obtained.

ATR is excellent for polymer films because the thickness of the film is not important. A 10 micron thick film and a 1000 micron thick film of the same material will give the same spectrum because the light penetrates to the same depth in both samples. One does not need to cast a film or press and heat the sample to obtain a quality spectrum as with transmission methods of polymer sampling. Usually, polymer films are clamped to the ATR crystal, and in some cases there is a gauge to help reproduce the pressure on the sample. Knowing the amount of pressure put on a sample is crucial since depth of penetration will only be reproducible if sample/crystal contact is reproducible. An example of the ATR spectrum of a polymer film is the spectrum of a latex glove shown in Figure 4.27. Once the spectrum is obtained, the polymer film can be unclamped and removed from the crystal and used for other

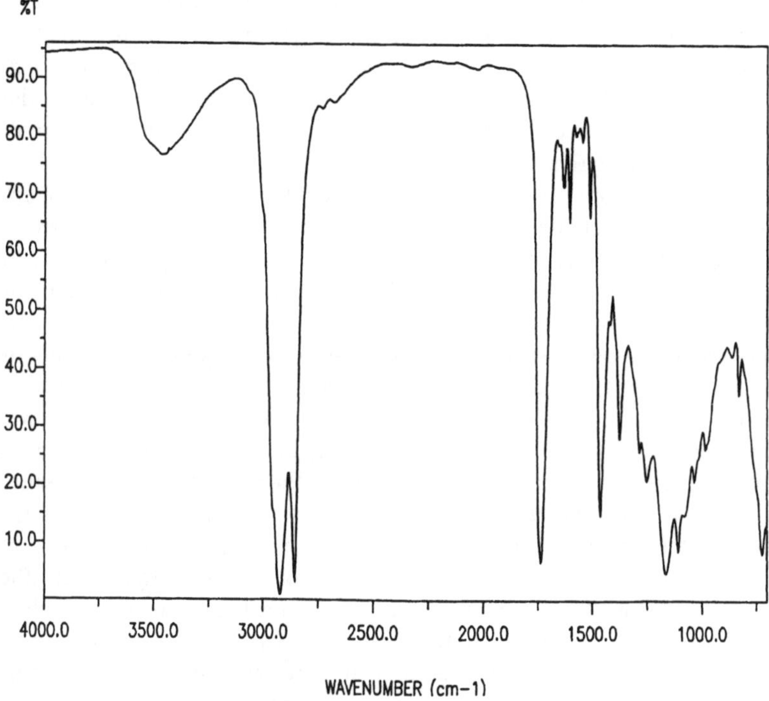

Figure 4.29 The ATR spectrum of a lipstick smear, illustrating how semi-solids can be sampled using this technique.

analyses. This points up another advantage of ATR, that it is a nondestructive sampling technique.

Another useful application of ATR is for soft or semisolid samples. A semi-solid is anything that is syrupy, viscous, gel-like, or a mixture of a liquid and a solid. Examples of these types of materials are shampoo, peanut butter, tomato sauce, toothpaste, motor oil, and asphalt (which is heated then smeared on the ATR crystal). The spectra of these samples can not be obtained in transmission because they are too absorbing, or are too viscous to form thin films. However, their ATR spectra can be obtained by simply smearing the sample on the ATR crystal. Since ATR crystals are not soluble in water, clean up is simply a matter of using water or an appropriate organic solvent to rinse the sample off the crystal. An example of the ATR spectrum of a semisolid is shown in Figure 4.29, which is the infrared spectrum of a lipstick smear. There are few other sampling techniques that can handle this type of sample as quickly and easily as ATR.

Liquid ATR is performed by simply suspending the ATR crystal in a vessel

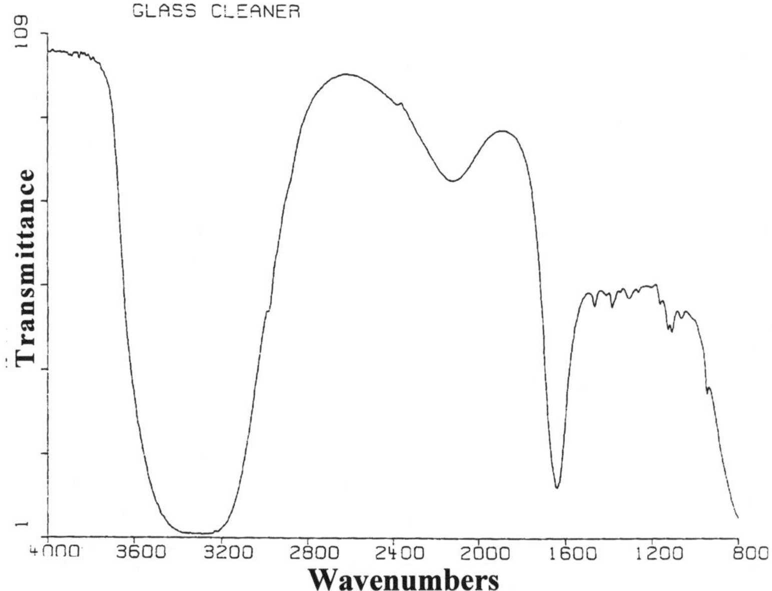

Figure 4.30 The liquid ATR spectrum of a glass cleaner. Note the intense water bands at 3500 and 1630 cm^{-1}.

that can hold liquids, and then pouring in the sample until it contacts the ATR crystal. Cleanup is simply a matter of pouring the sample out of the ATR cell, rinsing with an appropriate solvent, then allowing the crystal to dry. Liquid ATR is the sampling method of choice for aqueous solutions. ATR crystals are impervious to water, and the short pathlengths mean that water will not be totally absorbing. This allows water to be subtracted from sample spectra so that the spectra of solutes in aqueous solutions can be obtained. An example of how the spectrum of aqueous solutions can be obtained is shown in Figures 4.30 and 4.31. These figures show the spectra of a glass cleaner and the glass cleaner spectrum minus liquid water. Note in the subtraction result that the water bands have been totally removed, and that the glass cleaner spectrum has a good signal-to-noise ratio. A disadvantage of liquid ATR is that it is not very sensitive because the pathlengths are small. Solutes less than 0.1% in concentration cannot usually be detected.

A comment on the use of ATR for quantitative work is in order. As discussed above, depth of penetration can be easily controlled in ATR. However, the position and the condition of the ATR crystal can affect depth of penetration, and hence the results of a quantitative analysis. It is important to have a way of reproducibly positioning the crystal in the sampling accessory if quantitative analysis is going go be performed. Also, the crystal must be kept clean and scratch free. Scratches on an ATR crystal affect sample/crys-

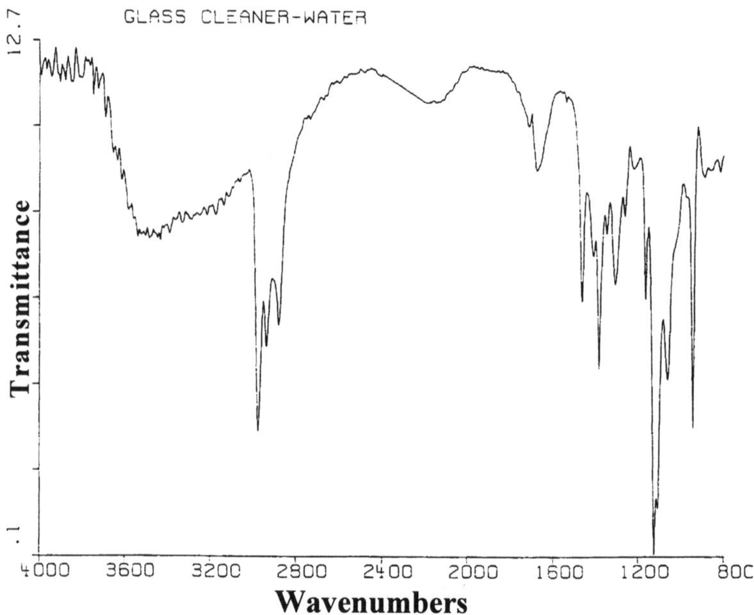

Figure 4.31 The subtraction result obtained after subtracting the spectrum of pure water from the spectrum of the glass cleaner seen in Figure 4.30. Note the absence of water bands.

tal contact and thus influence depth of penetration. Also, the same ATR crystal must be used for standards and samples when doing quantitative analysis. This does not mean any crystal of the right material can be used, but the exact same crystal must be used for standards and unknowns. If the crystal must be changed for any reason, new standards must be run and a new calibration performed before a quantitative analysis can be done.

The choice of an ATR crystal is very important. Table 4.1 lists the characteristics of the most common crystals. The color of each crystal is listed so you can tell which one you have. KRS-5 is the trade name for thallium iodide/thallium bromide. This material is extremely toxic, and should only be handled wearing gloves. It is a useful material because its transmission range covers the entire mid-infrared. Also, its low refractive index gives it the highest depth of penetration amongst the materials listed. The disadvantage of KRS-5 is that it is a soft material that can be easily scratched and bent. Care must be taken when cleaning the crystal to keep from scratching it. Zinc selenide (ZnSe) is the most common ATR crystal material. It is hard, impervious to everything but strong acids and bases, and has a depth of penetration similar to that of KRS-5. The only disadvantage of ZnSe is that it cuts off at

Table 4.1
Properties of Common ATR Crystal Materials

Material	$N_{crystal}$	Range (cm^{-1})	Color
KRS-5	2.35	20,000 - 250	Red
ZnSe	2.42	20,000 - 600	Yellow
Si	3.42	8300 - 660	Grey
Ge	4.0	5500 - 600	Gray

about 600 cm^{-1}, masking a small part of the mid-infrared from 600 to 400 cm^{-1}. Silicon (Si) and germanium (Ge) are also very hard, and are not attacked even by strong acids. They are usually employed when very small penetration depths are desired. Germanium is particularly useful for studying rubbery materials. These materials are filled with carbon black, which tends to scatter and absorb a lot of infrared radiation leading to curved baselines. By sampling the very surface of a material using a Ge crystal, the scattering is minimized, and features due to the rubber itself can be observed.

ATR is useful on solids and liquids, and certain ATR accessories have been designed to easily obtain the spectra of both kinds of materials. Thus, almost every sample you encounter can be studied using an ATR accessory.

Specular Reflectance

As mentioned above, *specular reflectance* takes place when the angle of incidence equals the angle of reflectance. Smooth surfaces such as mirrors give rise to specular reflectance. The way we see ourselves in a mirror is that light is reflected off our faces, is specularly reflected off the mirror, then enters our eyes. For infrared sampling purposes, there are really two different phenomena that fall under the heading of specular reflectance. These are illustrated in Figure 4.32. What is normally called specular reflectance is really first surface reflectance off a smooth substrate. If a film or coating of some sort exists on a smooth surface, the light beam passes through the coating, reflects off the smooth substrate, and passes through the coating a second time. This phenomenon is called *double transmission* or *reflection-absorption.* For our purposes, we will use the term double transmission.

Specular reflectance spectra are obtained with accessories that mount in the baseplate or slide into the sample slide mount in an FTIR's sample compartment. The optical diagram of a simple specular reflectance accessory is shown in Figure 4.33. It consists of two flat mirrors and a platform with a hole in it. The sample is placed on the platform. The first mirror directs the light to the sample, while the second mirror captures the light reflected off the sample and directs it to the detector. In theory, the infrared spectrum of anything that fits onto the platform can be obtained. In practice, this type of accessory is usually used to examine polymer films on metals. Background spectra are

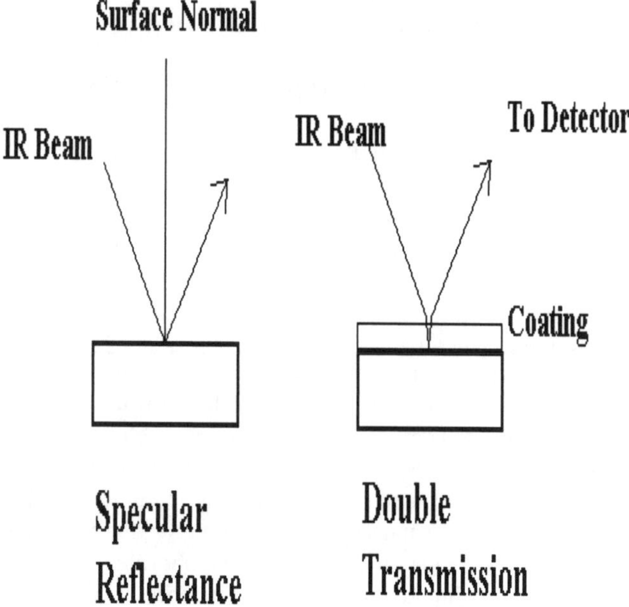

Figure 4.32 Examples of specular reflectance and double transmission. The latter technique is used to obtain the spectra of thin coatings on smooth surfaces.

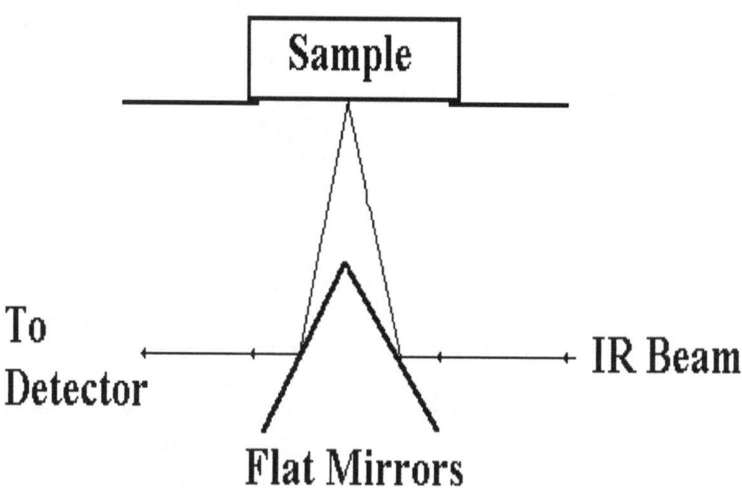

Figure 4.33 The optical diagram of a simple specular reflectance accessory.

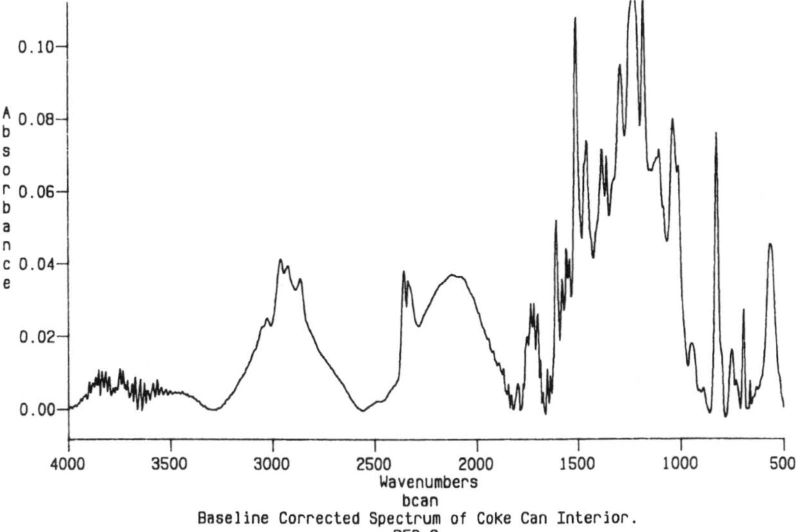

Figure 4.34 The specular reflectance spectrum of the inside coating of a Coke™ can. This spectrum has been baseline corrected.

obtained off "ideal reflectors" such as gold or aluminum mirrors. If your sample is a piece of coated metal, try and obtain the background spectrum with a piece of the same metal without the coating. This helps match the reflective properties of the sample and background and will produce a better spectrum. If the reflective properties of the background and sample are significantly different, a curved baseline may be seen in the sample spectrum representing the reflectivity differences between the sample and background. A double transmission infrared spectrum is shown in Figure 4.34. It shows the spectrum of the inside lining of a Coke™ can. The spectrum was baseline corrected because the reflective properties of the gold mirror used in the background spectrum and the metal on which the sample was coated were different enough to cause a curved baseline.

The advantage of the specular reflectance technique is that it is easy to perform, and involves little sample preparation. The problem with specular reflectance is that its applications are limited. If a sample does not have a smooth surface it will not give a good specular reflectance spectrum.

When true specular reflectance takes place off the first surface of a sample, the resultant spectrum may look very strange and contain derivative shaped bands. This is illustrated in Figure 4.35, which is the specular reflectance spectrum of a block of Plexiglass. The derivative shaped bands seen in the spectrum are known as *reststrahlen.* These bands are due to the fact that the reflectivity spectrum of a sample is dependent upon the behavior of the refractive index of the material with wavenumber. The refractive index of a material consists of real and imaginary parts as follows:

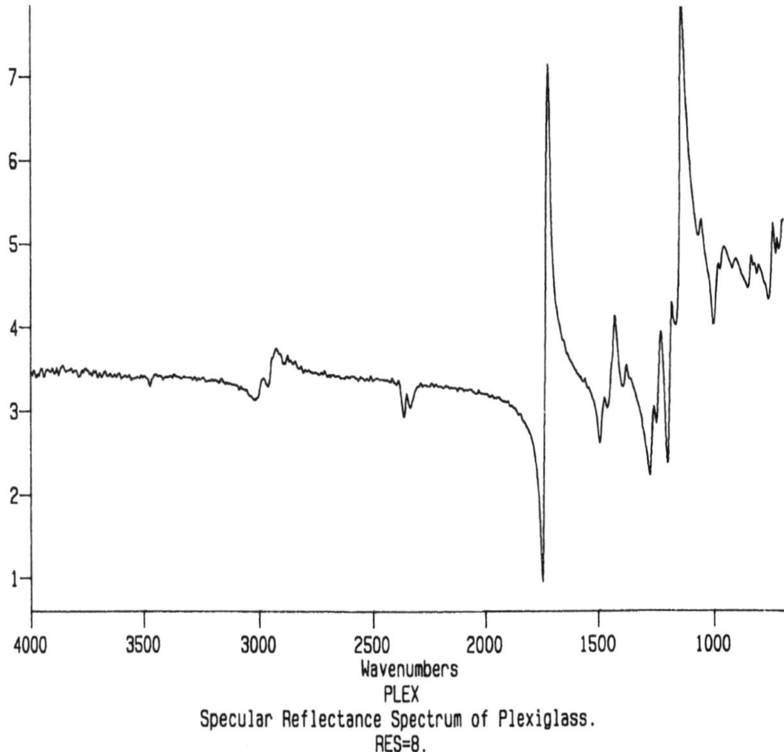

Figure 4.35 The specular reflectance spectrum of a Plexiglass block. The derivative shaped bands are called reststrahlen.

$$N = n + ik \qquad (4.6)$$

where:

N = The complex refractive index of a material
n = Real refractive index
k = Absorption coefficient
$i = (-1)^{1/2}$

The quantity N is said to be a complex number since it consists of a real part, n, and an imaginary part i (the square root of -1). Now, the real refractive index (n) of a substance changes drastically at wavenumbers where the sample absorbs radiation. If a plot of n is made versus wavenumber, a derivative shaped feature would appear at wavenumbers where the sample absorbs. A plot of k versus wavenumber is identical to the absorbance spectrum of a

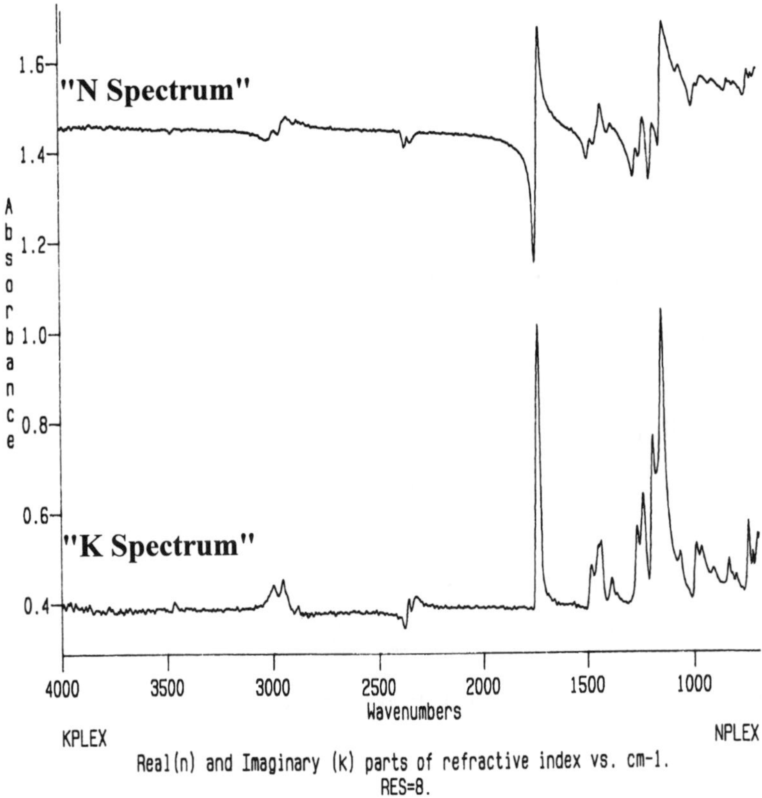

Figure 4.36 The n and k spectra calculated using a Kramers-Kronig transform on the spectrum shown in Figure 4.35.

sample. The reflectance spectrum is the sum of the variation of n and k with wavenumber. The derivative shapes seen in Figure 4.35 are due to the change of n with wavenumber, and this change is superimposed on top of the regular absorbance band. It is very difficult to interpret spectra such as the one in Figure 4.35. Fortunately, there is a way to perform a calculation on reflectance spectra to obtain separate plots of n and k versus wavenumber. This calculation is called a Kramers-Kronig (K-K) transform [4] (which has nothing to do with the Fourier transform discussed in Chapter 2). The results of performing a Kramers-Kronig transform on the spectrum in Figure 4.35 are shown in Figure 4.36. The top spectrum in the figure is a plot of n versus wavenumber, and is called the "n spectrum". The bottom spectrum in Figure 4.36 is a plot of k versus wavenumber, is equivalent to the absorbance spectrum of the sample, and is known as the "k spectrum". Many FTIR manufacturers sell special software that will perform K-K transforms on spectra. If a

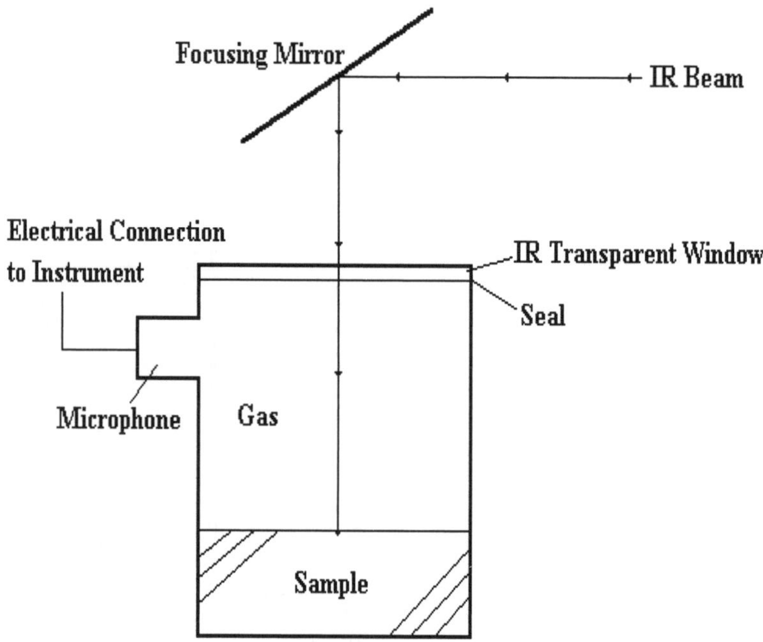

Figure 4.37 A diagram of a photoacoustic accessory.

sample has a somewhat rough surface, it may give rise to diffuse and specular reflectance. In this case performing a K-K transform may lead to distorted k spectra. So, these software packages work best on spectra that are mostly or purely specular.

C. Photoacoustic Spectroscopy (PAS)

Photoacoustic Spectroscopy (PAS) detects light absorption (photo) by using sound (acoustic). The photoacoustic effect was first described by Alexander Graham Bell over 100 years ago [5], but has been used to obtain spectra only in the last two decades. A diagram of a photoacoustic accessory is shown in Figure 4.37. Photoacoustic accessories drop into the sample compartment of most FTIRs. Samples are placed in a sample cup (about 1/2 inch across and 1/2 inch deep), and an infrared transparent window (usually KBr) is sealed above the sample to trap some gas between the sample and the window. Usually, helium is the gas used, although air may also be used. The PAS accessory is equipped with optics to direct the beam through the IR transparent window and focus the infrared radiation on the sample.

When infrared radiation is absorbed by a sample, it heats up. Different layers of the sample heat to different extents, depending on the concentration of absorbing species in the layer, and how much infrared radiation penetrates to that layer. Thermodynamics tells us that heat will diffuse from areas of high temperature to areas of low temperature. As a result, the heat from different layers in the sample diffuses towards the sample's surface. This movement of heat is known as a *thermal wave.* Once a thermal wave reaches the surface of the sample, the gas above the sample is heated, causing it to expand. This expansion causes a pressure wave to propagate through the gas, literally, the wind blows inside the sample cell. This wave would be detected by your ear as a sound, since sound is transported from its source to your ears by the movement of air molecules. The PAS accessory contains a sensitive microphone (originally developed for hearing aids) to hear the pressure wave. The microphone acts as a transducer, turning the sound into an electrical signal. There is an electrical connection between the microphone and the electronics on board the FTIR. A plot of the microphone signal versus optical path difference is an interferogram, which is Fourier transformed like any other interferogram to give an infrared spectrum. An example of a PAS spectrum is seen in Figure 4.38. Note that the peaks point up, and that the appearance of the spectrum is similar to an absorbance spectrum.

An advantage of the PAS technique is that it can, in theory, work on any solid, liquid, or gas that can be placed in the sample cell. This means that PAS is a universal technique. Another advantage of PAS is that it is nondestructive, and after obtaining a spectrum the sample can be removed from the sample cell and used for other things. A final advantage of PAS is that there is little sample preparation, simple place the sample in the accessory and start taking spectra. A disadvantage of PAS is that it is not very sensitive, and may give lower SNRs compared to other sampling techniques. However, there are certain samples for which PAS is the method of choice, as will be discussed below.

Photoacoustic spectroscopy is unique in that the accessory also contains the detector, so that there is no need for the infrared beam to leave the sample compartment. The material used in PAS to obtain background spectra is carbon black. Normally, one avoids taking infrared spectra of carbon black because it has broad, featureless absorbance bands that are not very useful. However, since carbon black absorbs at all infrared wavenumbers, it has a photoacoustic signal at all wavenumbers making it useful as a background material.

A key question in photoacoustic spectroscopy is from what depth in the sample does the signal arise. The sampling depth in a PAS experiment is defined as the depth from which 63% of the thermal wave that reaches the surface originates. The equation used to calculate the sampling depth is

$$L = (D/\pi F)^{1/2} \quad (4.7)$$

where:

L = Sampling depth in cm
D = Sample thermal diffusivity, in cm^2/sec
F = Frequency of modulation of radiation

When one needs to know the sampling depth in a PAS experiment, this equation can be used to calculate the value. Now, the frequency with which infrared radiation is modulated by an interferometer is given as follows (see Chapter 2):

$$F = 2vw \quad (4.8)$$

where:

F = Frequency of modulation
v = Moving mirror velocity, in cm/sec
w = Wavenumber of infrared radiation

By substituting equation 4.8 into equation 4.7 and rearranging, we obtain

$$L = (D/2\pi vw)^{1/2} \quad (4.9)$$

Note that the sampling depth is dependent on the wavenumber of infrared radiation in question, and that the sampling depth decreases as wavenumber increases. In this sense PAS is similar to ATR, where sampling depth also depends on wavenumber. Another important consequence of equation 4.9 is that the sampling depth is dependent on the moving mirror speed of the interferometer. This is a parameter that is easily varied on most spectrometers, so it is possible to obtain spectra from different depths within the same sample by simply changing the scan speed. The ability of PAS to obtain spectra from different depths within a sample is known as *depth profiling*. Depth profiling can produce interesting results if a sample's composition changes with depth. Now, it is important to realize that PAS cannot isolate the spectrum from a layer within a sample (unlike NMR imaging). Instead, when L is increased, the entire depth from which the signal is obtained increases. Thus, when L is small, the spectrum of the top layer of the sample is obtained. When L is large, the top layer and lower layers contribute to the infrared spectrum.

There are a number of phenomena that effect the intensity of a PAS signal. Some of the infrared radiation is reflected at the surface of the sample, and is lost. The thermal wave is attenuated to some extent by passing through the sample, and some of it is reflected at the sample's surface back into the sample, never to be detected. Despite all the different processes that contribute (or take away from) the intensity of a PAS signal, there is a linear relationship between the intensity of a PAS signal and the concentration of absorbing species in a sample. Thus, standards can be run, calibration curves plotted, and quantitative analyses performed as with other sampling techniques. The only

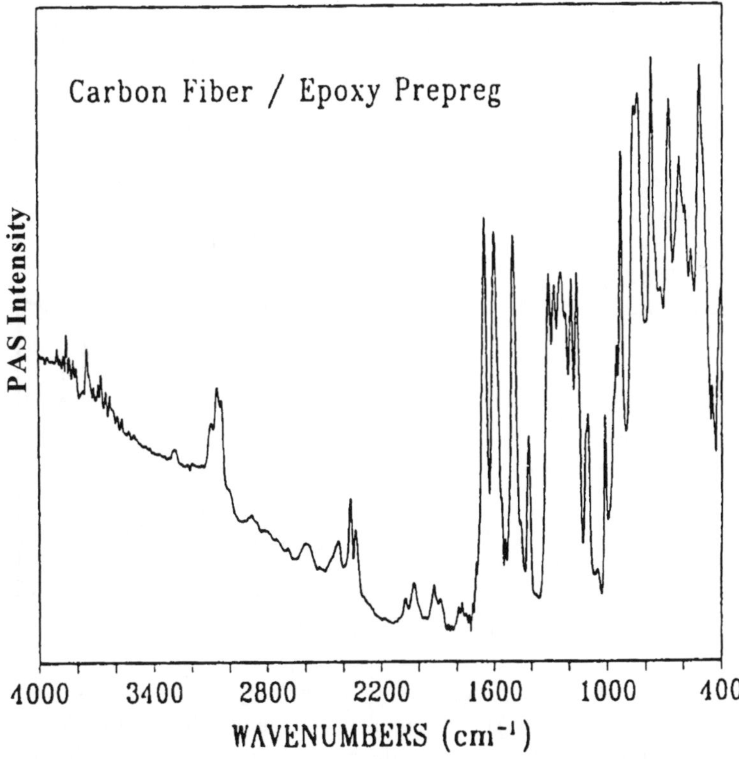

Figure 4.38 The photoacoustic spectrum of a carbon fiber/epoxy prepreg, which is used in the manufacture of carbon fiber materials. (Spectrum reprinted courtesy of MTEC Photoacoustics.)

caveat here is to use absorption bands that are not too intense. Intense PAS bands can undergo saturation, and changes in concentration will not result in linear changes in peak intensity.

PAS Applications

As mentioned above, photoacoustic spectroscopy is a universal sampling technique. However, there are several kinds of samples for which PAS is particularly well suited, and those samples will be emphasized in the following discussion.

Imagine trying to obtain the spectrum of a small piece of foam rubber. The material won't grind or press due to its elasticity, and therefore cannot be sampled using DRIFTS or KBr pellets. Foam rubber is cross-linked, so it is not soluble in solvents. This rules out casting a film. Lastly, foam rubber will not squish flat, so good contact cannot be made with an ATR crystal. Photoa-

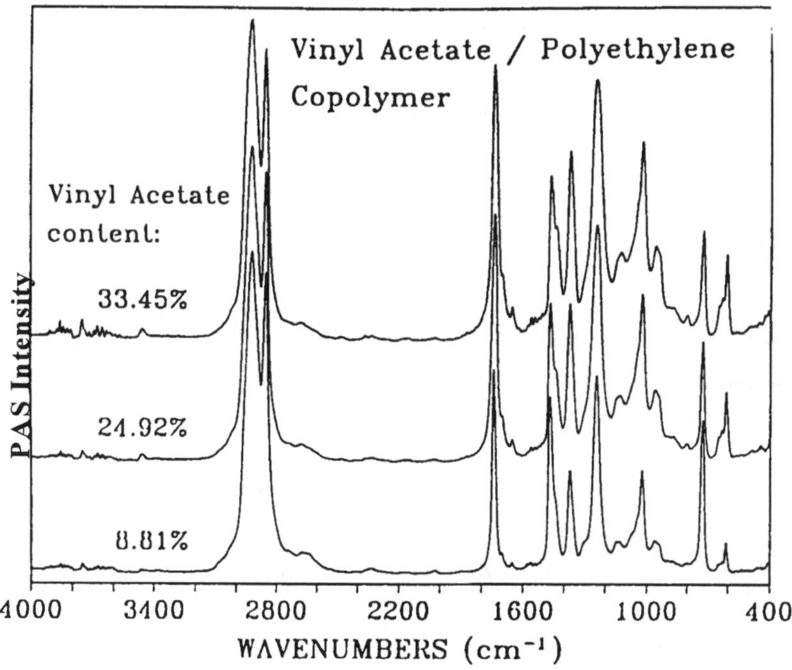

Figure 4.39 The PAS spectrum of a vinyl acetate/polyethylene copolymer in the form of a small pellet. The different spectra are of pellets with differing amounts of vinyl acetate. Several of the bands vary in intensity with vinyl acetate concentration, which means these spectra can be used for quantitative analysis. (Spectrum reprinted courtesy of MTEC Photoacoustics.)

coustic spectroscopy suffers from none of these problems. A small piece of foam rubber can simply be dropped into the sample cup of a PAS accessory, and its spectrum obtained.

Samples that contain large amounts of carbon black, such as rubber and carbon fiber composites, are difficult to analyze using traditional infrared sampling techniques. These samples tend to have broad featureless bands due to the carbon black, and can scatter much of the infrared radiation. By adjusting the moving mirror velocity, and thus the sampling depth, the spectrum of the top surface of these samples can be obtained, minimizing the carbon black bands. PAS uses carbon black as a reference material, and its contribution to the infrared spectrum of a sample can be ratioed out to obtain the spectra of other components in the sample, or of functional groups within the carbon black. This is illustrated in Figure 4.38, which is the PAS spectrum of a carbon fiber/epoxy prepreg. These samples are viscous mixtures containing epoxy resins and carbon fibers that are difficult to sample by any other tech-

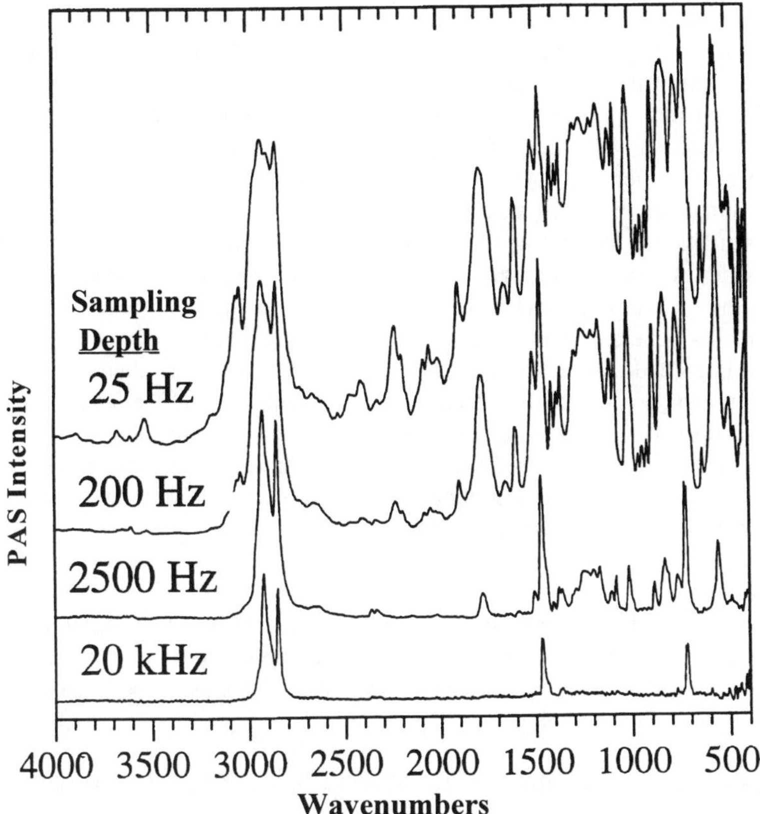

Figure 4.40 Photoacoustic depth profiling spectra of a thin polyethylene film coated on top of a polycarbonate. The bottom spectrum was taken at the fastest moving mirror speed, and hence has the smallest sampling depth. Reading up the figure, spectra were taken at slower speeds, and hence, larger sampling depths.

nique.

Small (1-3 mm diameter) polymer pellets are challenging infrared samples. These could be dissolved and cast as films, or heated and pressed to turn them into a film. However, this can be time consuming and alter the morphology and composition of the polymer. Also, some polymers simply do not dissolve or melt easily. The PAS spectrum of these samples can be obtained with no sample preparation. The pellet is simply placed in the sample cup. Figure 4.39 shows the PAS spectrum of a polyethylene/vinyl acetate copolymer pellet.

As mentioned above, PAS can be used for quantitative analysis. The three spectra shown in Figure 4.39 are of polymer pellets that contain varying amounts of vinyl acetate. Bands near 1700, 1370, 1200, and 1000 cm^{-1} due to the acetate moiety show intensity variations with changes in acetate content. Factor analysis (see Chapter 5) gave quantitative results for the acetate content accurate on the order of 0.5% [6]. Considering the ease with which these PAS spectra were obtained, it appears PAS is a good technique for quantitative and qualitative analysis of polymer pellets.

As mentioned above, PAS can be used to obtain depth profiling information on samples. An example is shown in Figure 4.40, which shows PAS spectra of a thin film of polyethylene coated on a polycarbonate. The bottom spectrum was obtained with the fastest scan speed, and hence had the smallest sampling depth. Each spectrum above the bottom one was taken at slower scan speeds, so the sampling depth increases as one reads up the figure. Note that at the smallest sampling depths, only the polyethylene bands are seen. As sampling depth is increased the polycarbonate and polyethylene bands are visible. Eventually, the sampling depth is so great that the polycarbonate bands mask the polyethylene bands.

Epilogue

This chapter has only scratched the surface of what is known about infrared sampling techniques. If more detail is desired on the techniques discussed here, or if knowledge of other techniques is desired, the user is referred to the books in the bibliography.

Companies Specializing in the Manufacture of FTIR Accessories

The following list of FTIR accessory manufacturers is provided as a public service, and is by no means complete. The inclusion or exclusion of any company in this list is not intended as an endorsement or a slight in any way.

CIC Photonics
2715 Broadbent Pkwy. NE
Albuquerque, NM 87107
(800) 635-3051
General FTIR Accessories

Harrick Scientific Corp.
88 Broadway
Ossining, NY 10562
(800) 248-3847
General FTIR Accessories

Janos Technology Corp
Rt. 35
Townshend, VT 05353
(802) 365-7714
General FTIR Accessories

Graesby/Specac
301 Commerce Dr.
Fairfield, CT 06430
(800) 447-2558
General FTIR Accessories

Spectra-Tech Inc.
2 Research Dr.
Shelton, CT 06484
(800) THE-FTIR
General FTIR Accessories

Carver Inc.
P.O. Box 544
Wabash, IN 46992
(219) 563-7577
Polymer and KBr Pellet Presses

Axiom Analytical
18103-C Sky Park South
Irvine, CA 92714
(714) 757-9300
Process Control FTIR Accessories

MTEC Photoacoustics Inc.
P.O. Box 1095
Ames, IA 50010
(515) 292-7974
Photoacoustic Accessories

Pike Technologies
2919 Commerce Park Dr.
Madison, WI 53719
(608) 274-2721
General FTIR Accessories

High Pressure Diamond Optics
231 W. Giaconda Way
Tucson, AZ 85704
(602) 544-9338
Diamond Anvil Cells

References

1. R. Silverstein, G. Bassler, T. Morrill, *Spectrometric Identification of Organic Compounds*, Wiley, New York, 1981.
2. M. Fuller and P. Griffiths, *Anal. Chem.* **50**(1978)1906.
3. P. Kubelka and F. Munk, *Z. Tech. Phys.* **12**(1931)593.
4. R. de L. Kronig and H.A. Kramers, *Atti Congr. Intern. Fisici, Como* **2**(1927)545.
5. A. G. Bell, *Proc. Am. Assoc. Adv. Sci.* **29**(1880)115.
6. J. McClelland, R. Jones, S. Luo, and L. Seaverson, A Practical Guide to FTIR Photoacoustic Spectroscopy, available from MTEC Photoacoustics, Ames IA, 1992.

Bibliography

P. Coleman, Ed., *Practical Sampling Techniques for Infrared Spectroscopy*, CRC Press, Boca Raton, 1993.

P. Griffiths and J. de Haseth, *Fourier Transform Infrared Spectrometry*, Wiley, New York, 1986.

N. Colthup, L. Daly, and S. Wiberley, *Introduction to Infrared and Raman Spectroscopy*, Academic Press, New York, 1990.

N.J. Harrick, *Optical Spectroscopy: Sampling Techniques Manual*, Harrick Scientific, Ossining, NY, 1987.

J. Robinson, *Undergraduate Instrumental Analysis*, Marcel Dekker, New York, 1995.

D. Skoog and D. West, *Principles of Instrumental Analysis*, Holt, Rinehart & Winston, New York, 1971.

Chapter 5

Quantitative Analysis

A. Introduction

The purpose of infrared quantitative analysis is to determine the concentration of a molecule or molecules in a sample. The molecule of interest is called the *analyte*. The height or area of a peak in an absorbance spectrum is directly proportional to the concentration of analyte in the infrared beam. To establish the correlation between absorbance and concentration, the infrared spectra of samples that contain known concentrations of the analyte are measured. These samples are called *standards*. During a process called *calibration;* a mathematical model is generated giving the correlation between absorbance and analyte concentration. Once an accurate model is in hand, the concentrations of analytes in unknown samples can be predicted. All quantitative analyses assume that the model describes the unknown samples as accurately as the standards. This is generally true. However, much of the work involved in obtaining a successful quantitative analysis is done to insure the model is realistic and accurate. Thus, it is very important to pay attention to detail and reproducibility when performing a quantitative analysis. Infrared spectroscopy can be used to quantify from one to many components in a sample.

The first part of this chapter covers Beer's law, standard methods, and how to perform single component quantitative analyses. The next part describes 15 experimental pitfalls of quantitative analysis, and how best to avoid them. The last section of the chapter focuses on multicomponent quantitative analysis, i.e., how to determine the concentration of more than one molecule at a time in sample. This process is somewhat mathematically complex. However, the discussion tries to explain the math in an easy to understand manner, and emphasizes the good and bad points of the different mathematical techniques used in multicomponent quantitative analysis. The chapter ends with a brief overview of factor analysis, which is a new and exciting way of determining multiple concentrations in a sample.

B. Beer's Law

The basis of all quantitative analyses is Beer's law, which relates concentration to absorbance. It has the following form:

$$A = \varepsilon \, l \, c \quad (5.1)$$

where:

A = Absorbance
ε = Absorptivity
l = Pathlength
c = Concentration

For liquids the pathlength is usually given in microns, and the concentration in moles /liter. You can use whatever units for pathlength and concentration that are convenient, as long as you are consistent in their use. The absorptivity is the proportionality constant between concentration and absorbance. It changes from wavenumber to wavenumber for a given molecule, and is different for different molecules. However, for a given molecule and wavenumber of absorbance, the absorptivity is a fundamental physical property of the molecule. For example, the absorptivity of acetone at 1700 cm^{-1} is different than the absorptivity for acetone at 1690 cm^{-1}. However, the absorptivity of acetone at 1700 cm^{-1} is a fundamental property of the molecule as invariant as the molecule's boiling point or molecular weight. The units of ε are usually given in (concentration x pathlength)$^{-1}$, so the absorptivity cancels the units of the other two variables in Beer's law. This is necessary because absorbance is a unitless quantity. The absorbance is measured as a peak height, peak height ratio, peak area, or peak area ratio from the FTIR spectrum. An example of an absorbance band is seen in Figure 5.1.

Traditionally, infrared spectra have been plotted with the Y axis units as either absorbance, where the peaks point up, or in transmittance, where the peaks point down. For qualitative work it doesn't matter what units you use, since peak position on the X axis is the thing of interest. However, for quantitative analysis it is absolutely necessary that you use infrared spectra plotted in absorbance. This is because absorbance is linearly proportional to concentration as stated in Beer's law. Transmittance is not linearly proportional to concentration, and it is defined in the following equation:

$$T = I/I_0 \quad (5.2)$$

where:

T = Transmittance
I = Light intensity with a sample in the beam
I_0 = Light intensity with no sample in the beam

In FTIR, I_0 is the background spectrum, and I is the sample single beam spec-

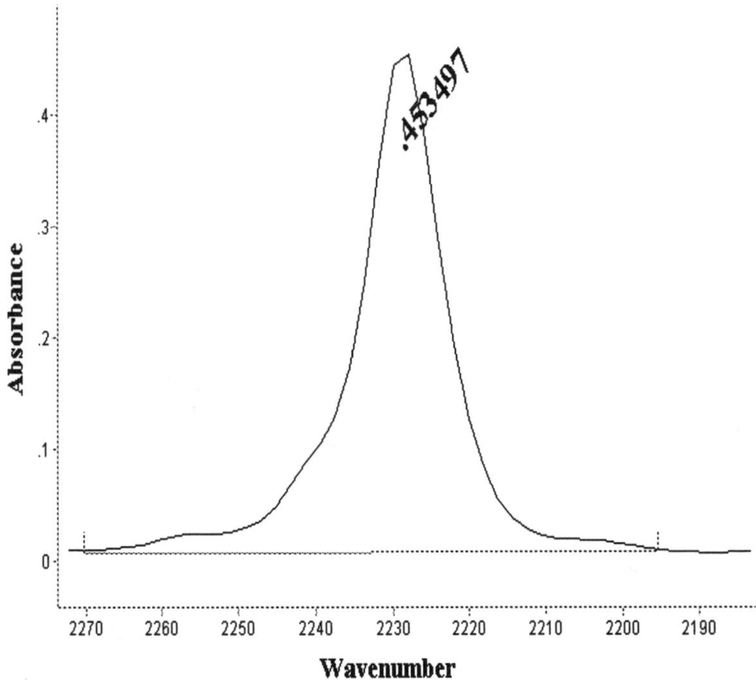

Figure 5.1 An example of an absorbance peak. The absorbance value (peak height) for this peak at 2228 cm^{-1} is 0.453. The integrated peak area calculated using the baseline and peak edges shown could also be used as the absorbance value.

trum. Absorbance is defined as

$$A = \log I_0/I \quad (5.3)$$

where "log" is the symbol for the base 10 logarithm of a number. Rearranging equation 5.2 shows that

$$1/T = I_0/I \quad (5.4)$$

Substituting equation 5.4 into 5.3 yields

$$A = \log(1/T) \quad (5.5)$$

Equation 5.5 establishes the relationship between absorbance and transmittance. Most FTIR software packages allow you to switch between the two

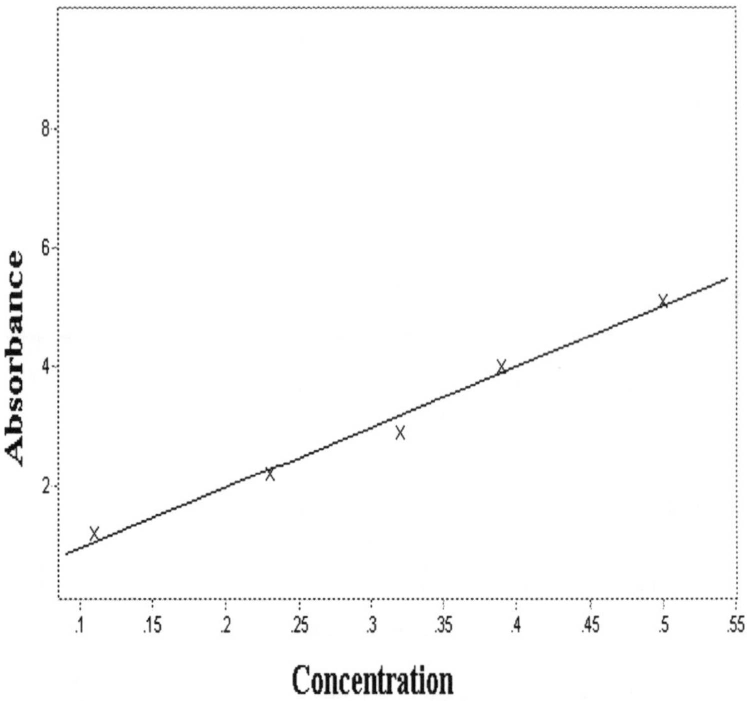

Figure 5.2 A plot of absorbance versus concentration for a single analyte, called a calibration curve.

units, and equation 5.5 is what the computer uses to make the conversion. If we substitute Beer's law (Eq. 5.1) into equation 5.5 we obtain:

$$\varepsilon lc = \log(1/T)$$

Raising both sides of the equation to the power of 10 yields

$$10^{\varepsilon lc} = 1/T$$

and rearranging yields the final result.

$$T = 10^{-\varepsilon lc} \quad (5.6)$$

Equation 5.6 shows that the relationship between transmittance and concentration is that of an inverse logarithm. This is not a linear function, making transmittance spectra inappropriate for use in quantitative analysis. Only absorbance, Kubelka-Munk, and photoacoustic spectra (the latter two types of spectra also have peak heights that are linearly proportional to concentration) should be used to perform quantitative analyses.

To perform a calibration using Beer's law, we must first observe that the form of Beer's law is analogous to the formula for a straight line. The formula for a straight line is

$$Y = mx + b \quad (5.7)$$

where:

Y = Y axis value of data point
m = Slope of line
X = X axis value of data point
b = Y-intercept of the line (where the line crosses the Y axis)

Comparing equations 5.1 and 5.7 shows that a plot with absorbance on the Y axis vs. concentration on the X axis should give a straight line with a slope of εl and a Y intercept of zero. For a single component quantitative analysis, a calibration is performed by measuring the absorbances of a series of standards. The absorbance of the standards at a specific wavenumber, or integrated peak area is plotted vs. concentration. If this plot is linear, then Beer's law is obeyed by the system, and the straight line is called a *calibration curve* (it is called a curve even though the plot is a straight line). An example of a calibration curve is shown in Figure 5.2. For single component analysis, the calibration curve is the mathematical model of the standards.

The slope of the calibration curve can be measured, as for any straight line, by choosing two data points on the line and calculating the difference between their Y axis values (ΔY), then calculating the difference between their X axis values (ΔX), then dividing $\Delta Y/\Delta X$ (or put more simply, calculate rise/run). By rearranging Beer's law and applying equation 5.7 we obtain the following:

$$C_{unknown} = A/\varepsilon l$$

$$\text{Slope} = \varepsilon l$$

$$C_{unknown} = A/\text{Slope} \quad (5.8)$$

where:

$C_{unknown}$ = The concentration of analyte in an unknown sample

Predicting the concentration of an analyte in an unknown sample is simply a matter of dividing the measured absorbance of the unknown sample by the slope of the calibration curve.

If the calibration curve is not linear, Beer's law is not obeyed, and quantitative analysis should not be attempted. There are a number of reasons why Beer's law may not be obeyed. For example, as the concentration of a molecule in a sample is increased, chemical interactions may affect the spectrum of the molecule to the point that Beer's law is not followed. This is why intense absorbance bands are rarely used in quantitative analysis. Another reason Beer's law may not be followed is due to chemical reactions within the sample. These reactions may cause the concentrations of the analyte to change in an unpredictable fashion, causing the calibration curve to be nonlinear. Finally, instrumental problems can cause a system to not follow Beer's law. A detailed discussion of the instrumental limitations of quantitative analysis in FTIR is given in the book by Griffiths and de Haseth listed in the Bibliography.

As mentioned above, the calibration curve is really a mathematical model of the standard samples. The curve gives the response of the instrument to changes in analyte concentration. It is assumed that the unknown samples are similar enough to the standards that the absorbances measured will follow the calibration curve. This is why it is vital that the standard samples be as similar as possible to the unknowns. Also, it is vital that standards and unknown samples be run using identical experimental conditions. The only thing that should cause the instrument's response to change from sample to sample is changes in the concentration of the analyte.

C. Single Component Analyses

Performing a single component analysis is the simplest application of Beer's law. The following six steps outline the steps to follow to assure accurate single component analyses.

<u>Analyzing For a Single Component</u>

1. Prepare the standards. Be as accurate as possible since any concentration error in the standards is carried through to generate error in the predicted concentration of the unknown samples.
2. Obtain the spectra of the standards in random order. Run the spectrum of each standard twice.
3. Subtract the two spectra of the same standard using a subtraction factor of one. A straight line should be the result. If not, something changed between running the two spectra. Examine your instrument, the sample, and the sampling technique to determine where the source of error occurred. At a minimum, rerun the spectra of the standard.
4. Examine the spectra and choose a band (or bands) whose height or area changes with concentration.
5. Measure the appropriate absorbance(s) and plot the calibration curve (see

above). Examine the curve to insure linearity, then calculate its slope.
6. Obtain the spectra of the unknown samples, and calculate the unknown concentrations using the calibration curve's slope and the absorbance of the unknown (equation 5.8).

If these simple steps are followed, single component quantitative analysis should be fairly straightforward.

As mentioned above, a calibration curve can be plotted using peak heights, peak areas, peak height ratios, and peak area ratios. An example of an absorbance peak is shown in Figure 5.1. Peak areas are calculated by simply adding together all the absorbances in a specific wavenumber range. Now, the sign of random error can be positive or negative. When data points are added together, the positive and negative noise contributions will cancel, improving the quality of the data (this is the reason adding scans together and smoothing reduces noise levels, as discussed in Chapters 2). This is why peak areas are better suited for quantitative analysis than peak heights.

If you accurately know the pathlength of the sample, then using peak areas or peak heights is fine. However, if the pathlength of the samples is not known, it can be taken care of by ratioing the peak heights or areas measured at different wavenumbers in the same spectrum as follows:

$$A_1 = \varepsilon_1 lc$$

$$A_2 = \varepsilon_2 lc$$

where:

A_1 = Absorbance at wavenumber 1
A_2 = Absorbance at wavenumber 2

When these two absorbances are ratioed we obtain the following:

$$A_1/A_2 = \varepsilon_1 lc/\varepsilon_2 lc$$

The pathlength is the same in the numerator and in the denominator, so it cancels and the result is a pathlength independent quantity. When plotting a calibration curve using peak height or area ratios, the ratio is used as the absorbance value on the Y axis. The ratio from an unknown spectrum must be calculated in the same way as for the standards, then the unknown concentration can be predicted. When using sampling techniques such as KBr pellets and DRIFTS, where it is difficult to know the pathlength of the sample, the use of peak height or peak area ratios is recommended.

D. Different Standard Methods

There are two ways of using standards to perform a calibration, which will be discussed below. It will be assumed that only one analyte concentration is being measured.

External Standards

The method of *external standards* works as follows. Standard samples containing known but different concentrations of the analyte are made. The infrared spectra of the standards are obtained, and their absorbances are measured. The peak heights or areas are plotted against concentration to give a calibration curve (see Figure 5.2). The concentration of an unknown sample is determined by obtaining its infrared spectrum, measuring its peak height or area, and then using the calibration curve to predict the concentration in the unknown sample. This is called the method of external standards because the standards and the unknown samples are run at different points in time.

The external standards method works fine for many samples, and is relatively simple to perform. However, it does have some drawbacks. It does not take into account fluctuations in instrument performance, sampling, or other random errors that may occur between when the standards are analyzed and when the unknown samples are analyzed. These variables can have an unknown effect on the measured absorbance, contributing error to the final concentration determination. One can never be 100% sure if the calibration performed at some other point in time works now for the instrument and sample you are using working with.

Internal Standards

The method of *internal standards* involves addition of a known amount of a known material to each standard, and to each unknown sample. This material, known as the *internal standard*, should be chemically stable, not interact with other molecules in the sample, and have a unique infrared peak. When running standards by this method the absorbances (peak height or area) of the analyte and the internal standard are ratioed. The calibration curve is generated by plotting this ratio on the Y axis versus the concentration of analyte on the X axis. For unknown samples, the ratio of the analyte and internal standard absorbances is calculated, then used with the calibration curve to predict the concentration of analyte in the unknown. This method assumes that any unknown source of error, such as sampling or instrument performance, will affect the internal standard and analyte absorbances equally. When these absorbances are ratioed the effects of the unknown variables are cancelled, yielding a more accurate analysis. As a result, internal standards analyses are normally more accurate than external standards analyses. However, the internal standards method involves more work and calculations than the external standards method.

E. Helpful Hints/Experimental Pitfalls to Avoid

Following is a list of common experimental problems that can ruin your quantitative analyses, and some helpful hints you can follow to insure that your analyses are always accurate and reproducible.

1. Use standards with concentrations that bracket the expected concentration range of components in your unknowns. Concentrations measured outside the range used in the standards are invalid. For example, if your standards run from 10% to 90% in component A, only unknowns whose concentration falls between 10% and 90% component A can be analyzed. Your mathematical model cannot be applied to concentrations greater than 90% or less than 10% because you have no information on the behavior of the model in those concentration ranges. Therefore, if any predicted concentration falls outside the range used in your standards, ignore them.

2. In preparing standards for multicomponent analysis, make up standards whose concentrations are linearly independent of each other. That is, the concentrations of successive standards should not be simple multiples of each other, or add up to a constant. Otherwise, independent information on each standard will not be available.

3. Use the actual components to be measured when preparing standards. For instance, if you are analyzing for technical grade acetone (95% pure) do not use reagent grade acetone (99% pure) in your standards. Remember that the calibration curve is a mathematical model of your unknowns, and for the model to be accurate, your standards must be as similar to the unknowns as possible.

4. Run standards in random order to ensure sample order has no effect on the results. Obtain two spectra of each standard, and subtract them using a subtraction factor of 1.0 to see if a straight line is obtained. If a straight line is not obtained, it means that a variable in the sample or the experiment has caused spectra of the same sample to be different. The source of variability must be found and eliminated before attempting to perform a quantitative analysis.

5. Be careful when making standards. The accuracy of the final results can be no better than the accuracy to which the standards were prepared. If the concentrations in your standards are accurate to 1%, no amount of careful sample preparation, calibration curve plotting, or statistical analysis will make the final results any more accurate than 1%. The accuracy with which you know the concentrations in the standards is the fundamental limit on the accuracy of a predicted concentration. There are a number of things that can make the accuracy worse than this number, but there is nothing you can do to make the accuracy better than this number.

6. The minimum number of standards to run is 2n+2, where n is the number of components to be analyzed. Thus, for a one component system the number of standards to run is ((2x1)+2) or 4. By overdetermining the problem, the effects of random variables are minimized.

7. Cleanliness is crucial! Thoroughly clean all cells, crystals, windows, etc.

that your sample comes in contact with before performing an analysis. Occasionally obtain the spectrum of sample cells, ATR crystals, and KBr windows by ratioing their single beam spectra against the spectrum of air to check for contamination.

8. Only use absorbance bands that are less than 0.8 absorbance units. Beer's law is usually not obeyed by bands whose absorbances are greater than this. If necessary, dilute the sample or change the pathlength to get the absorbance to less than 0.8.

9. Be aware of the chemistry of your sample and the sampling device. Water and samples containing water are not compatible with KBr. Copper complexes, strong acids, and strong bases have been known to attack ZnSe ATR crystals. Grind KBr and samples separately to avoid reactions under pressure between the KBr and the sample. Make sure your sample components do not react with each other. This can be done by obtaining spectra of the sample over time and looking for changes in the spectrum.

10. The physical condition of the standards and unknowns, e.g., temperature, and pressure must be the same when their spectra are acquired. Daily fluctuations in room temperature and pressure should not be a problem. However, do not run your standards at room temperature, then an unknown sample at 150°C. The spectra will be too different to obtain an accurate quantitative analysis.

11. Use all the same instrumental parameters, e.g., number of scans, resolution, wavenumber range, and apodization function for both the standards and unknowns. You don't want changes in the instrument contributing to changes in your results. Many FTIR software packages allow you to save a set of instrumental parameters on the computer's hard disk. By saving the parameters used when the standards are run, then loading the same parameter set when unknowns are run, you guarantee that the same parameter set is used for both kinds of samples.

12. Do as little spectral manipulation as possible on standards and unknowns. Baseline correction and smoothing should be avoided. If you must manipulate your data before quantitation, all standards and unknowns must be manipulated in exactly the same fashion. Only experienced FTIR users should attempt quantitation after spectral manipulation.

13. Be consistent in the use of cells, windows, crystals, and sampling accessories for your standards and unknowns. A calibration developed using a ZnSe ATR crystal will not work for a Ge crystal. If a new crystal, cell, window, etc. is used, check your calibration (see #14).

14. Once your calibration curve is established, and your analysis is working properly, occasionally check your calibration by running a new standard and comparing the known and calculated concentrations. If the concentrations are similar, the calibration curve is still good. If the concentrations do not agree, then changes in the sample, sampling accessory, and instrument over time may have contributed to drift in quantitative results. You must track down the source of variability, eliminate it, and then run a new set of standards and plot a new calibration curve. Daily or weekly checks of calibration can ensure accurate results, and contribute to the peace of mind of the analyst.

15. Technique specific problems regarding quantitative analysis are discussed in the sampling techniques section. Please reread the section appropriate for your sampling technique before running any samples.

Despite the problems discussed here, thousands of quantitative FTIR analyses are performed each day, and quantitative analyses can be made routine. The point to remember is that if you develop the method properly using the information contained above, you will have fewer problems later when the method is actually being used.

F. Multicomponent Quantitative Analysis

Multicomponent quantitative analysis is used to measure the concentrations of several components at a time in complex mixtures. Although much work must be done obtaining spectra of standards and choosing the wavenumbers to use, the concentrations of many components in a mixture can be accurately determined from one spectrum of the unknown sample. The basis of all quantitative analyses is Beer's law, which is described above. The following discussion assumes familiarity with Beer's law, single component analysis, and the experimental details of performing quantitative analyses described earlier in this chapter.

All multicomponent analyses are based on the additivity of Beer's law, i.e., the absorbance at a specific wavenumber is the sum of the absorbance of all sample components which absorb at that wavenumber. This idea is expressed in equation form as follows. For a three component system, where the three components are denoted by the subscripts d, e, and f, at a given wavenumber the total absorbance is given by:

$$A_t = A_d + A_e + A_f$$

where:

A_t = The total absorbance at a given wavenumber

$A_{d,e,f}$ = The absorbances of component d, e, or f at a given wavenumber

If several molecules absorb infrared radiation at the same wavenumber, the absorbance at that wavenumber is simply equal to the sum of the absorbances of the individual molecules. Now, each molecule follows Beer's law, and by applying equation 5.1 we obtain the following:

$$A_t = \varepsilon_d l c_d + \varepsilon_e l c_e + \varepsilon_f l c_f$$

where:

$\varepsilon_{d,e,f}$ = The absorptivity of component d, e, or f
l = Pathlength
$c_{d,e,f}$ = The concentration of component d, e, or f

Now, we can not solve for the concentration of three components with just one equation. We need three equations to solve for three unknowns. This is accomplished by measuring the absorbance at three different wavenumbers, and setting up and solving a system of three equations in three unknowns. The equations are as follows (components are denoted by subscripts d, e, and f; the three different wavenumbers are denoted by subscripts 1, 2, and 3):

$$A_1 = e_{1d} l c_d + e_{1e} l c_e + e_{1f} l c_f$$
$$A_2 = e_{2d} l c_d + e_{2e} l c_e + e_{2f} l c_f$$
$$A_3 = e_{3d} l c_d + e_{3e} l c_e + e_{3f} l c_f$$

Systems of equations such as the above can be rewritten in matrix form. A matrix is simply a group of numbers. These groups have specific properties, and can be manipulated mathematically. For a review of matrix algebra, any college level calculus or linear algebra book should suffice. If a matrix consists of just one column, it is called a vector. The system of equations shown above for the three concentrations can be expressed in matrix form as follows:

$$\mathbf{A} = \mathbf{E}\ \mathbf{L}\ \mathbf{C} \quad (5.9)$$

where:

A = The vector of absorbances
E = The matrix absorptivities
L = The vector of pathlengths
C = The vector of concentrations

From now on, vectors and matrices will be denoted by symbols in **bold text**. Note that ε changes with wavenumber and component. All the different mathematical methods used in multicomponent quantitative analysis start with a matrix equation such as equation 5.9. Calibrations are performed by obtaining spectra of standards containing known concentrations of the components of interest. The standards supply the information for **A**, **L**, and **C**; the calibration provides the values for **E**. The different multicomponent quantitative analysis techniques perform matrix algebra upon the data in different ways to generate a calibration. Once the calibration is known, predicting the concentrations of unknowns is simply a matter of measuring the absorbances of the sample and entering them into the calibration.

The following discussion will focus on the mathematical differences between multicomponent quantitative techniques, and also emphasize the practical differences between the different methods such as speed of analysis, accuracy, and the number of standards needed. The types of mathematical techniques available to you will be dependent on the software you have. In general, the calculations described below are transparent to the user when using a computer, and the choice of which technique to use is simply a matter of setting a software parameter. However, an understanding of what the computer is doing to your data is necessary to truly understand how to develop an accurate quantitative method. Be aware that the same high quality lab technique is necessary for accurate multicomponent analysis as well as accurate single component analyses. No amount of matrix manipulation or statistical analysis can compensate for poorly made standards, sloppy technique, or lousy spectra.

The K and P Matrix Methods

The K matrix method, also known as the Classical Least Squares (CLS) method, combines the pathlength and absorptivity matrices into a single matrix called K as follows:

$$\mathbf{K} = \mathbf{E}\,\mathbf{L}$$

As a result, equation 5.9 is rewritten as:

$$\mathbf{A} = \mathbf{K}\,\mathbf{C}$$

Note that the absorbance is written as a function of concentration. To obtain a calibration, the absorbance and concentration information obtained from the standards is used to obtain the values in the K matrix as follows:

$$\mathbf{K} = \mathbf{A}\,\mathbf{C}^{T}\,(\mathbf{C}\,\mathbf{C}^{T})^{-1} \qquad (5.10)$$

where $\mathbf{C}^{T}$ is the transpose of matrix **C**, and the superscript $^{-1}$ stands for the inverse of a matrix. A matrix transpose involves swapping the rows and the columns of a matrix. A matrix inverse is best thought of as a form of matrix division. The matrix operation shown in equation 5.10 is known as a *least squares fit*, and by definition will produce the best available model of the data. Once **K** is known, the unknown concentrations comprising the matrix **C** can be calculated as follows:

$$\mathbf{C} = (\mathbf{K}^{T}\,\mathbf{K})^{-1}\,\mathbf{K}^{T}\,\mathbf{A}$$

Again, this is a least squares fit, and will produce the best possible prediction of **C**. Note that by calculating **C** all the unknown concentrations are determined at once.

The K matrix method is useful because the mathematics are straightforward, and many standards and wavenumbers can be used in the calibration to obtain an averaging effect. The problem with the K matrix method is that it is very sensitive to the presence of impurities or unexpected components in a sample. It is also intolerant of interactions between sample components. For the K matrix method to work properly, the concentration of every chemical species in the standards must be known. It is difficult to know the complete chemical composition of complex samples, as impurities are always present. This lack of knowledge can greatly increase the error in a K matrix quantitation. The mathematical reason for this can be understood by looking at the above equations. In the K matrix method, absorbance is expressed as a function of concentration. If an impurity is ignored in the calibration, it means an entire column is missing in the C matrix, and the matrix multiplications needed for calibration cannot be performed properly. The K matrix technique is best used in situations where the sample composition is predictable, and there is no interaction between the components.

Fortunately, there are other ways of formulating the above matrix calculations to avoid the sensitivity of the calculation to unknown impurities. Equation 5.9 can be rewritten as follows:

$$\mathbf{C} = \mathbf{P}\,\mathbf{A}$$

where:

$$\mathbf{P} = (\mathbf{E}\,\mathbf{L})^{-1}$$

Quantitative analyses performed using this method are known as P matrix analyses, or Inverse Least Squares (ILS) analyses. Note that in this method concentration is written as a function of absorbance, whereas in the K matrix method absorbance is written as a function of concentration. This may seem like a trivial distinction, but in practice it proves to be very important. If the concentration of an impurity is ignored in the P matrix method, it means the C matrix is missing a column, but as long as there are as many absorbance measurements as there are concentrations, the matrix multiplication needed to obtain P can be performed. Therefore, only the concentrations of the components of interest need be known to perform a P matrix analysis. The P matrix is calculated to obtain a calibration as follows:

$$\mathbf{P} = \mathbf{C}\,\mathbf{A}^{\mathrm{T}}\,(\mathbf{A}\,\mathbf{A}^{\mathrm{T}})^{-1}$$

This is another least squares calculation, and the prediction of unknown concentrations is performed using the same least squares matrix manipulations as for the K matrix method discussed above.

The drawback of the P matrix method is that there must be as many absorbance measurements as components in the sample, and the absorbances selected must be sensitive to changes in the component's concentration. If the number of components is unknown, many absorbance measurements must be made to ensure the number of absorbances is greater than the number of components. Another drawback of the P matrix method is you must use as many standards as their are absorbance measurements. This means many more standards must be run than in K matrix analyses. This also means that not every wavenumber in a spectrum can be used in a quantitative analysis, unless several thousand standards are run! Thus, one is not making optimal use of all the spectral data available. So, the real disadvantages of the P matrix method is that it involves running more standards and measuring more absorbances than the K matrix method, and it makes inefficient use of the available data. However, since it is difficult to always account for every component in a sample, P matrix methods are almost always the matrix method of choice in multicomponent quantitative analyses.

Factor Analysis: PCR and PLS

Many successful analyses have been performed using the K and P matrix methods outlined above. However, the disadvantages of these approaches has lead to research into other multicomponent quantitative methods. A group of techniques known as *factor analysis* methods avoid the disadvantages of the K and P matrix methods. The best known types of factor analysis are called Principal Components Regression (PCR) and Partial Least Squares (PLS). Factor analysis is insensitive to the presence of impurities, makes full use of all spectral data, and does not require a standard for each wavenumber used. As such, factor analysis combines the advantages of the matrix methods with few of the disadvantages. In addition, factor analysis will help identify the number of components in a mixture, reproduce the spectrum of each component, and predict the component concentrations in unknown samples. Finally, factor analysis is tolerant of low quality data. Spectra that are noisy, obtained at low resolution, or that contain artifacts can still be successfully used in factor analysis. The K and P matrix methods do not run successfully using low quality data.

All this may seem too good to be true, and in a sense it is. The disadvantage of factor analysis is that the mathematics are difficult to understand, and some skill is involved in performing the analyses properly. Their are several different factor analysis methods, but PCR is the simplest to understand. It works as follows. First, one must obtain the spectra of standards, and enter concentration information about the standards into the software. Second, the part of the spectrum to be used must be chosen. There is no real limit here; one can use the entire spectrum if desired. However, the analysis will take longer if more data points are chosen. PCR performs a calibration as follows. The average of the standard spectra is calculated. This average spectrum is compared to each standard spectrum, and a new spectrum, called the *factor,* describing the differences between the standards and the average is calculated (hence the term factor analysis). The amount of the factor in each standard spectrum is calculated, and is called a *score*. There is a score for each standard spectrum. Next, the score for each standard spectrum is multiplied by the factor, and this is subtracted from each standard spectrum to produce a residual. The residuals are used in the next iteration of the technique to calculate a new average spectrum, a new factor, and a new set of scores. The process is repeated over and over, and more factors and scores are generated until the user stops the process.

Another way of describing factor analysis is to say that the data set undergoes spectral decomposition. The first factor is related to the spectrum of the component whose concentration varies the most among the standards. The first set of scores are related to the concentration of this component in each standard. When the first factor is subtracted from all the standard spectra, the contribution of this component to the spectra is removed. The second factor is related to the spectrum of the component whose concentration varies by the second largest amount, and so on. With each iteration, the spectral contribution of some component is stripped from the standard spectra. Factor analysis is sensitive to the variability in the standard spectra, and uses factors and scores to describe that variability. However, a factor can describe noise just as well as concentration variations. Thus, one of the jobs of the user is to determine the optimum number of factors.

The number of factors is related to the number of components, but the two numbers are not necessarily equal. If the number of factors is less than the number of components, the concentrations of two components may be related to each other, and hence can be described with one factor. If the number of factors is more than the number of components, it means there may be components present you do not know about, or some other source of variability such as sampling errors or instrument problems are contributing to the spectra. The scores in factor analysis are related to the concentration of the components in the different standards.

The factor analysis program does not know how many factors are needed to best describe the data. It will eventually start modeling noise, or other sources of variability. Hence, the skill required in using this technique is to choose the correct number of factors to model the concentration data accurately. Fortunately, there is help available. A process called *cross-validation* is used to test for the optimum number of factors. This is done by removing one standard spectrum from the data set, performing a calibration without the standard, then using this calibration to predict the composition of all the standards. The difference between the predicted and actual concentrations of the standards is calculated while the number of factors is varied. In turn, each standard spectrum is removed from the data set, a calibration is performed, and a set of concentrations predicted. The optimal number of factors will be the minimum number that does the best job of predicting the composition of the standards. There are graphical and numerical ways of displaying this information to make choosing the correct number of factors easy. Once the correct number of factors is chosen, a final calibration is performed, and the concentration of unknown samples can be predicted. This involves simply measuring the spectrum of the unknown sample, and applying the factor analysis to the spectrum.

The above discussion has centered around the PCR technique. The other factor analysis technique is known as the Partial Least Squares (PLS) method. It differs from PCR in that PLS uses the concentration information to calculate a weighted average spectrum. Thus, the factors and scores calculated in PLS are different from PCR. There are also two ways of performing PLS, known as PLS-1 and PLS-2. In PLS-1 a separate set of scores and factors is calculated for each component of interest, while PLS-2, like PCR, calculates one set of factors and scores for the entire set of data. Thus, PLS-1 is probably more accurate than PLS-2 in that scores and factors are tuned to each component. However, PLS-2 takes much less computer time. In general, PLS-1 works better on systems where the concentrations vary widely, while PLS-2 gives equivalent results on systems where the concentrations do not vary widely. In general, either PLS method is more accurate than the PCR method described above. PCR is better known because it is the easiest factor analysis technique to understand. The choice of factor analysis method is ultimately up to the analyst. A good quantitative analysis software package will allow you to choose different factor analysis methods, and to pick the one that obtains the best results.

Factor analysis is used in ways other than to predict the concentrations of chemicals in unknown samples. This technique is sensitive to any source of variation in a spectrum. As long as a physical or chemical property of a sample causes the sample's spectrum to change in some way, the spectral changes and measurements of the property can be correlated. For instance, things such as boiling point, viscosity, and octane number have been correlated with spectral changes using factor analysis. These properties can then be predicted for samples by simply obtaining their spectrum. In general, it is easier to obtain the infrared spectrum of a sample than measure many other physical properties. So, by correlating the infrared spectra with other sample properties, analysis time and dollars can be saved.

Bibliography

J. Duckworth, "PLS Plus Users Manual"; available from Galactic Industries, Salem N.H., 1990.

P. Griffiths and J. de Haseth, *Fourier Transform Infrared Spectrometry*, Wiley, New York, 1986.

W. George and H. Willies, Eds., *Computer Methods in UV, Visible, and IR Spectroscopy*, Royal Society of Chemistry, Cambridge U.K., 1990.

H. Freiser, *Concepts & Calculations in Analytical Chemistry,* CRC Press, Boca Raton FL., 1992.

S.V. Compton and D.A.C. Compton, in *Practical Sampling Techniques for Infrared Analysis,* P. Coleman, Ed., CRC Press, Boca Raton FL, 1993.

N. Colthup, L. Daly, and S. Wiberley, *Introduction to Infrared and Raman Spectroscopy*, Academic Press, Boston, 1990.

D. Shoemaker and C. Garland, *Experiments in Physical Chemistry,* McGraw-Hill, New York, 1967.

D. Skoog and D. West, *Principles of Instrumental Analysis,* Holt, Rinehart, & Winston, New York, 1971.

J. Robinson, *Undergraduate Instrumental Analysis*, Marcel Dekker, New York, 1995.

J. Ferraro and K. Krishnan, Eds., *Practical Fourier Transform Infrared Spectroscopy*, Academic Press, New York, 1990.

Chapter 6

Hyphenated Infrared Techniques

As the rest of this book has shown, infrared spectroscopy is an excellent tool for determining the identity and quantities of molecules in a sample. However, there are many other analytical techniques that measure the physical properties of a sample and provide little or no chemical information. As a result, many separate physical and chemical measurements may be necessary to fully characterize a sample. It would be nice if both physical and chemical information could be obtained on a sample at the same time using one instrument. By interfacing an FTIR with an instrument that measures a sample's physical properties, both chemical and physical information about a sample can be obtained at the same time This greatly increases the amount of information known about a sample, and decreases the amount of analysis time compared to obtaining the information via separate analyses. When FTIRs are interfaced with other pieces of equipment new techniques are invented. These are called "hyphenated infrared techniques" because the acronyms used to describe the techniques end in "-FTIR". For example, "GC-FTIR" is used to refer to Gas Chromatography-Fourier Transform Infrared Spectroscopy.

This chapter will focus on a discussion of several of the most common or interesting hyphenated infrared techniques. The first, infrared microscopy, involves interfacing a visible light microscope with an FTIR to obtain spectra of very small (10 to 250 microns in size) samples. In GC-FTIR a gas chromatograph separates volatile samples into their pure components before their spectra are obtained by an FTIR. Subsequent spectral interpretation and library searching can establish the identity of the molecules in a sample. Similarly, in High Pressure Liquid Chromatography-FTIR (LC-FTIR) a liquid chromatograph separates samples into their component molecules before the FTIR obtains their spectra.

A. Infrared Microscopy

Infrared microscopy involves coupling an FTIR to a visible light microscope, and allows for visual and infrared examination of microscopic samples.

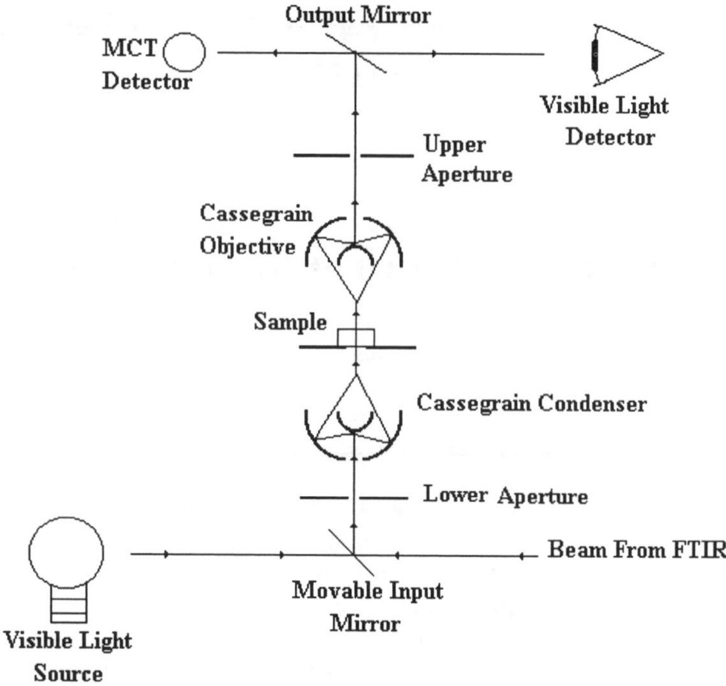

Figure 6.1 An optical diagram of an infrared microscope.

Infrared microscopes are often times high quality visible microscopes that have been redesigned for use with infrared radiation. A schematic diagram of a hypothetical infrared microscope is shown in Figure 6.1. This diagram represents a combination of several currently used microscope optical designs, and is not intended to represent any manufacturer's specific offering. Typically, the infrared microscope mounts on the side of the FTIR. The infrared beam is "picked off" by a mirror after leaving the interferometer and is directed through a hole in the side of the spectrometer into the microscope. This beam allows the IR spectrum of the sample to be obtained. There is also a visible light source (a light bulb) mounted in the microscope so the sample can be examined visually. A movable input mirror is used to choose whether infrared or visible light irradiates the sample. Thus, an infrared microscope can be used in visual mode or in infrared mode by irradiating the sample with either kind of light.

In some microscope designs, the infrared beam passes through a lower aperture, which usually consists of four black knife edges whose position can be adjusted by the user. The purpose of the lower aperture is to define the shape of the infrared beam prior to the sample. The optic used in a microscope to focus light on the sample is called a *condenser*. In many infrared microscopes the condenser consists of a mirror arrangement called a Cassegrainian optic (we will call it a Cassegrain for short), as seen in Figure 6.1. A hole in the bottom of the optic allows the infrared light through, which then bounces off of a small convex mirror. A larger concave mirror collects the infrared beam, and focuses it on the sample. Samples are typically placed on a microscope X,Y translation stage to allow different parts of the sample to be viewed. The details of sample preparation will be discussed below.

The optic used to collect light after it has interacted with the sample is called the *objective*. In most infrared microscopes this is another Cassegrain mirror identical to the condenser except that it has been flipped over and now points down rather than up. In the objective the concave mirror collects the infrared beam and focuses it on the convex mirror. The convex mirror reflects the beam straight up towards the detector.

After the objective there is an aperture which is used to define the area whose infrared spectrum is going to be obtained. For many samples, the size of the infrared beam at its focal point is larger than the sample itself. For example, the infrared beam may be 250 microns in diameter, but the sample may only be 100 microns in diameter. As a result, some of the infrared beam does not interact with the sample. This light is useless because it contains no information about the sample. Also, impurities in the vicinity of the sample may cause spurious bands to appear in the spectrum of the sample. The upper aperture prevents spurious bands by keeping stray light away from the detector. The size of the aperture is adjusted while in visible light mode. The image of the aperture's knife edges are brought into contact with the image of the area to be sampled, then the switch to infrared mode is made and the spectrum of the defined area is obtained.

After the upper aperture the infrared beam is brought to a focus by appropriate optics onto a detector. The infrared detector must be a high sensitivity, narrow band, liquid nitrogen cooled MCT detector. Use of a sensitive detector is essential since infrared microscopy is a "light-starved" technique, and very little infrared radiation makes it to the detector due to reflection losses and aperturing. In visible mode the optics are arranged to allow viewing of the sample by reflecting the beam to an eyepiece.

In infrared microscopy it is imperative that the sample and background spectra be run using the same aperture setting. This is as important as using the same number of scans and the same resolution for these spectra. Recall that the sample determines the aperture size in infrared microscopy, so the sample must be located and the aperture adjusted before taking a background spectrum. Now, microscopic samples are easily lost due to vibrations and wind currents. So, once a microscopic sample is in view in the microscope's eyepiece, and the aperture is set properly, the single beam sample spectrum should be obtained immediately. The background spectrum is obtained after the sample spectrum. The sample single beam spectrum and the background can be ratioed at a later time to obtain the absorbance or transmittance spectrum.

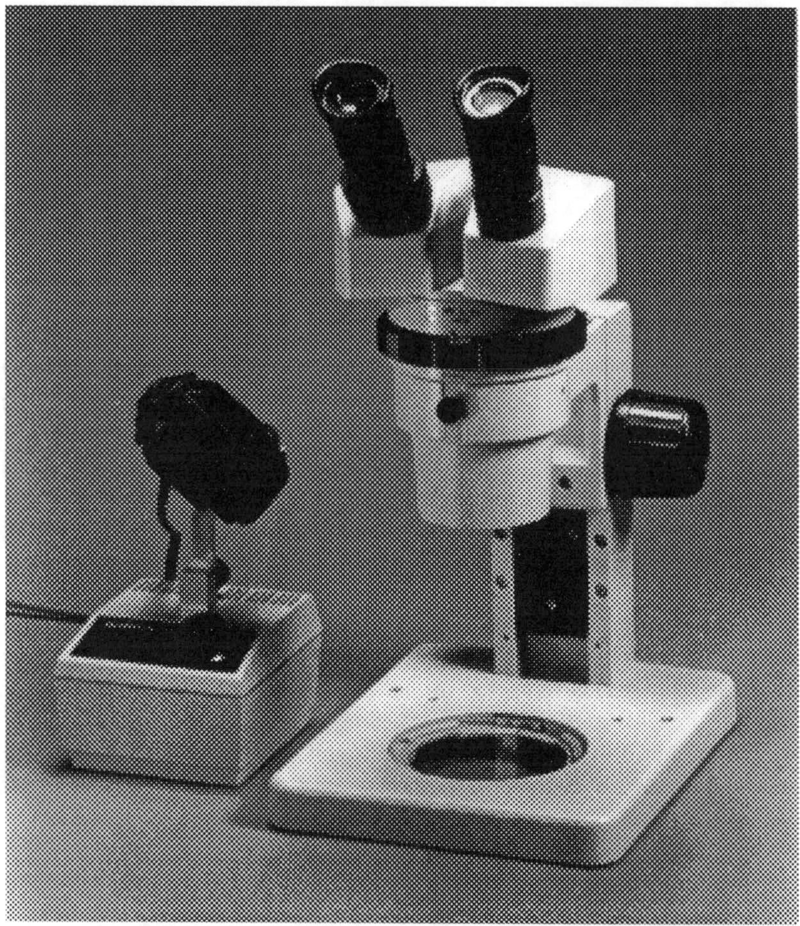

Figure 6.2 An example of a preparative microscope used to prepare samples for analysis by infrared microscopy. (Photo courtesy of Spectra-Tech.)

Figure 6.1 shows the optical configuration used when the light beam is transmitted by the sample. It is also possible to use an infrared microscope in reflectance. In this case the infrared radiation comes from above the sample and is focused using the upper Cassegrain. The light reflects off the sample and is collected by the upper Cassegrain, and is eventually focused on the detector. We have now reviewed the four ways in which an infrared microscope can be used: visual transmission, visual reflection, infrared transmission, and infrared reflection.

Sample Preparation

The manipulation and preparation of samples for infrared microscopy requires the use of a low power stereo microscope for sample preparation. A picture of such a microscope is shown in Figure 6.2. The sample is placed on the working surface of the microscope, and is illuminated with a lamp. Special tools such as probes, scalpels, and tweezers are used to cut the sample, move it around, and mount it for examination by the infrared microscope. It takes skill, patience, and practice to manipulate small samples while looking through a microscope. Steady hands are a must for anyone preparing microscopic samples (so it may be a good idea to not drink coffee before working with a microscope!). One way to steady your hands while working with these samples is to plant the little finger of each hand on the table top, and manipulate the sampling tools with the other fingers.

Microscopic samples can be analyzed in transmission or reflectance. Microscope transmission samples suffer from the same "thickness problem" as macroscopic samples (see Chapter 4). The thickness of the sample must be adjusted to between 1 and 20 microns to let the right amount of light through. Soft samples can be easily flattened with a press, or with a small roller that is part of the handle of a scalpel. These "roller knives" are available from your microscope manufacturer, and are effective because they simultaneously flatten and stretch the sample, quickly getting the sample to the right thickness. A wide variety of samples, including polymers, powders, and crystalline materials can be flattened with a roller knife.

Once the sample is flattened, it is mounted on an infrared transparent window (usually KBr). The window is placed on the infrared microscope's stage with the microscope in visual mode. Since the sample is much smaller than the window, it is sometimes difficult to locate. By placing an identifying scratch on the window near the sample, finding it is made easier. Once the sample is located the aperture(s) are adjusted, and the sample single beam spectrum is obtained. The sample is taken out of the field of view, and the background spectrum is taken through a clean part of the KBr window using the same aperture setting as the sample. Using the FTIR's software, the sample single beam spectrum must be ratioed to the background spectrum to obtain an absorbance or transmittance spectrum. The microscope transmittance spectrum of a crystal of cocaine is seen in Figure 6.3.

If a sample is too tough to be flattened with a roller knife, a *diamond anvil cell* can be used to flatten the sample. A photo of a diamond anvil cell is shown in Figure 6.4. The cell consists of two small gem quality diamonds with very flat faces that are mounted parallel to each other. The sample is placed on one diamond, and the second diamond is brought into contact with the first, flattening the sample in the process. Often times finger pressure is enough to flatten the sample. Otherwise, there are screws that can be tightened by hand to flatten the sample. The entire cell is placed in the infrared microscope, and the infrared beam is passed through the sample and the diamonds. The background spectrum must be acquired through a clean part of

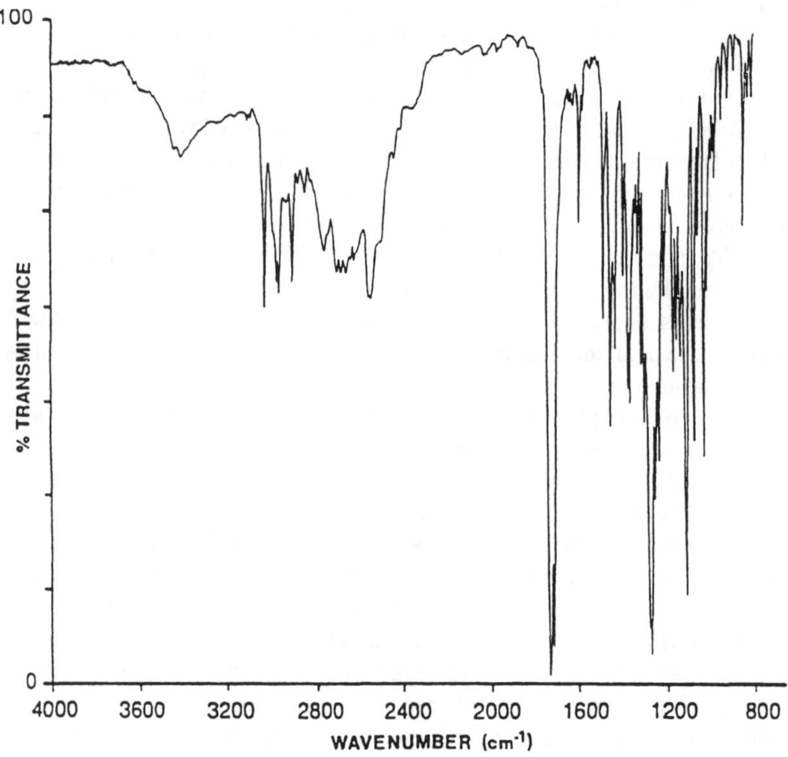

Figure 6.3 The infrared spectrum of a cocaine crystal obtained in transmission using an infrared microscope.

the diamond. This can be difficult if the sample covers most of the diamond's surface and there is not a clean area of diamond the same size as the sample. In these cases the cell must be removed from the microscope, cleaned, then put back on the microscope stage and the background spectrum obtained without changing the aperture.

Diamond has infrared bands around 2000 cm^{-1} which sometimes show up in the sample spectrum. A way to reduce the intensity of the diamond bands is to flatten the sample, take the diamond cell apart, mount the sample on one diamond, then take the infrared spectrum through the one diamond. Very few materials absorb around 2000 cm^{-1}, so the diamond bands are not a problem for most samples. Diamond anvil cells are available from several of the accessory companies listed at the end of Chapter 4.

Infrared spectra of microscopic samples can also be obtained in reflectance. The sample is typically mounted on a gold mirror, and the infrared beam is bounced off the sample. The background spectrum is obtained by reflecting

Figure 6.4 A photo of a diamand anvil cell, which is used to flatten samples for transmission analysis. (Photo courtesy of Spectra-Tech.)

the light off a clean portion of the gold mirror. Gold and aluminum mirrors are used for the background spectrum because they are excellent infrared reflectors.

Since reflectance techniques do not suffer from a thickness problem, it is not necessary to flatten reflectance samples, which saves a great deal of time. A disadvantage of the reflectance technique is that a lot of the infrared beam is scattered by the sample, is not collected by the Cassegrain, and never reaches the detector. This means reflectance spectra are usually noisier than transmission spectra. Of course, the number of scans can always be increased to achieve a reasonable signal-to-noise ratio. Microscope reflectance samples are one of the few times when upwards of 1000 scans may be needed to obtain a good SNR.

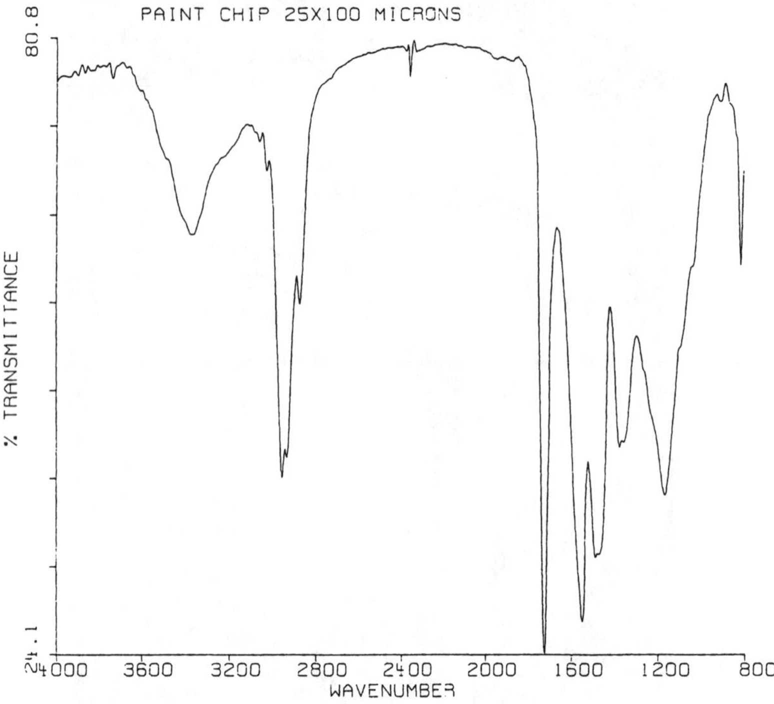

Figure 6.5 The infrared spectrum of a paint chip, obtained using an infrared microscope equipped with a diamond anvil cell.

Applications

The applications of infrared microscopy are widespread. In fact, many people buy FTIRs for the sole purpose of performing infrared microscopy. One of the most important applications of infrared microscopy is in forensic labs, i.e., police and crime labs. Items left at crime scenes can provide clues as to the perpetrator of a crime. Paint chips are often left behind on victims of hit and run automobile accidents. These chips are recovered from the victim's body, are pressed flat in a diamond anvil cell, and the microscope spectrum obtained in transmission. The spectrum of an automobile paint chip is seen in Figure 6.5. Automobile manufacturers and law enforcement officials have cooperated to maintain libraries of infrared spectra of paint chips from automobiles. By performing a library search, the make, model, and year of a car can be determined. A search of the Department of Motor Vehicles database can be made to find out who in the area owns the type of car involved in the accident. This information has been used in courts of law to obtain felony convictions.

Clothing fibers are also left behind at crime scenes by criminals. The infra-

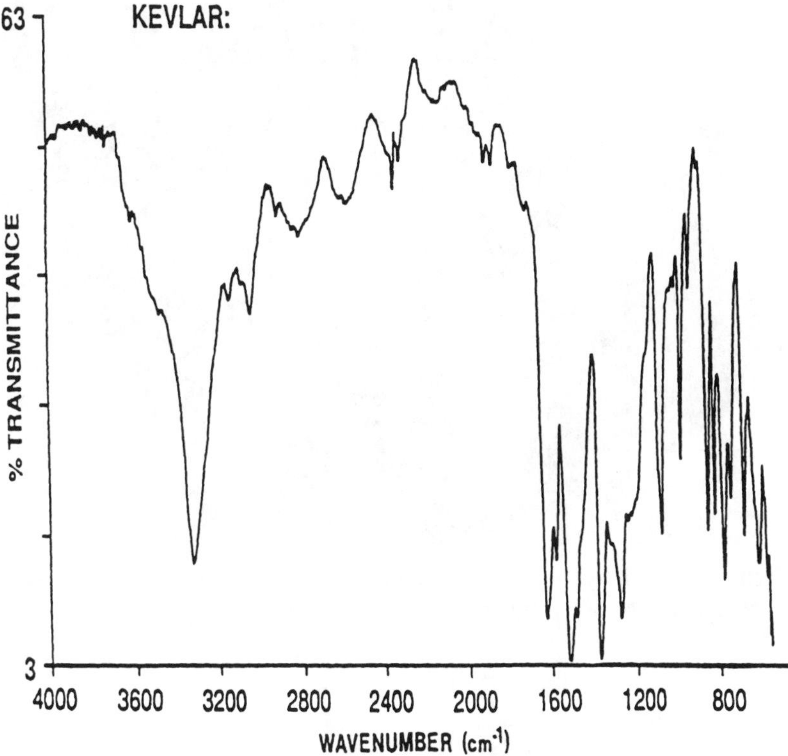

Figure 6.6 The infrared spectrum of a single Kevlar fiber, obtained with an infrared microscope.

red spectra of clothing fibers are obtained in transmission using a diamond anvil cell. By matching fibers from the crime scene with fibers from a suspect's clothing, circumstantial evidence as to the presence of the suspect at the crime scene can be provided. Clothing fiber evidence is not as informative as paint chip evidence since many articles of clothing are made of the same fibers, i.e., polyester and cotton. The infrared spectrum of a Kevlar single fiber, which is a polyamide, is shown in Figure 6.6.

Any product such as paper, polymer films, and coatings that exist as thin sheets can contain microscopic contaminants and defects. These defects can consist of foreign particles that create a bump in the sheet, or of areas where the sheet is too thin. These defects can render a product useless. A photograph of a defect in a polymer film is seen in Figure 6.7. Infrared microscopy is often times the tool used to analyze these defects. The offending portion of a sheet is cut out, and the reflectance spectrum of the defect is obtained. Once the defect is identified, the source of the contaminant may be found and elimi-

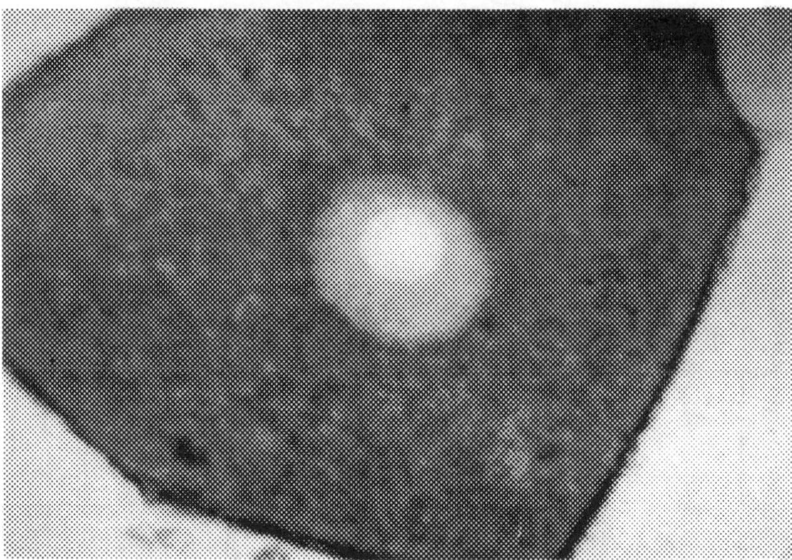

Figure 6.7 A photograph of a defect in a polymer. This defect was analyzed in reflectance using infrared microscope mapping.

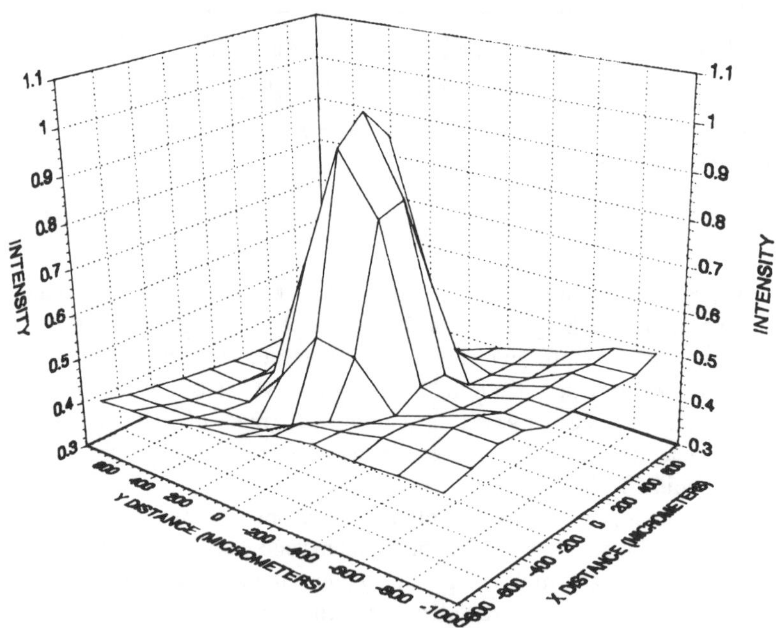

Figure 6.8 A microscope map of the polymer defect shown in Figure 6.7. This is a plot of the infrared absorbance at 1730 cm^{-1} vs. position. This indicates the defect was caused by plasticizer in the polymer.

nated.

Another way to study a sample such as a polymer defect is to use a technique known as *microscope mapping.* In this technique, a computer controlled microscope stage precisely positions different parts of the sample in the IR beam. In a typical experiment, spectra are obtained in a grid pattern and a spectrum is obtained at each point in the grid. For example, if the grid pattern is 500 microns square, and a spectrum is obtained every 50 microns along the grid, >100 spectra would have to be obtained to complete the pattern. The spectra are combined to give a plot of infrared absorbance at a specific wavenumber vs. position. These plots, known as *molecular maps,* show how the composition of a sample varies from place to place. An example of a molecular map is shown in Figure 6.8, which is a plot of the absorbance at 1730 cm^{-1} vs. position for the polymer defect seen in Figure 6.7. The absorbance at 1730 cm^{-1} is due to plasticizer in the polymer, and indicates the plasticizer was somehow involved in creating the defect.

An important application of infrared microscopy is the study of biological samples. Kidney stones, arteries, and individual cells have all been studied using this technique. An example of this is the spectrum of an individual cheek cell seen in Figure 6.9. The most exciting development in the field of biological infrared microscopy is the discovery that healthy and cancerous human cervical cells have different infrared spectra [1]. The day may come when infrared microscopes are used to help detect and cure cancer.

B. Gas Chromatography-Fourier Transform Infrared Spectroscopy (GC-FTIR)

Gas chromatography is a method used to analyze mixtures of volatile molecules. The sample is injected into a heated port where it is vaporized. A flow of an inert gas, usually helium, sweeps the sample into a silica column that resides in an oven. The column is lined with adsorbent material (typically some type of siloxane) upon which the sample molecules adsorb. The oven is heated at a predetermined rate, and different molecules deadsorb from the column at different temperatures. The temperature at which a molecule deadsorbs is determined by the boiling point of the molecule, and to a lesser extent by its molecular structure. Different molecules will come off the column at different points in time and are detected using a flame ionization, thermal conductivity, or element specific detector. The fundamental measurement taken with a GC is called a *chromatogram*, which is a plot of detector response vs. time. The peaks in a chromatogram are denoted by their *retention time*, which is a measure of how long after sample injection a peak comes off the column. The peak heights and areas in a chromatogram are proportional to the amount of substance coming off the column at that point in time. Quantitative analysis can be performed in this way. Gas chromatography is a widely used technique for analyzing the composition of complex mixtures. Its applications include use in environmental analysis, the petroleum industry, and chemical quality control.

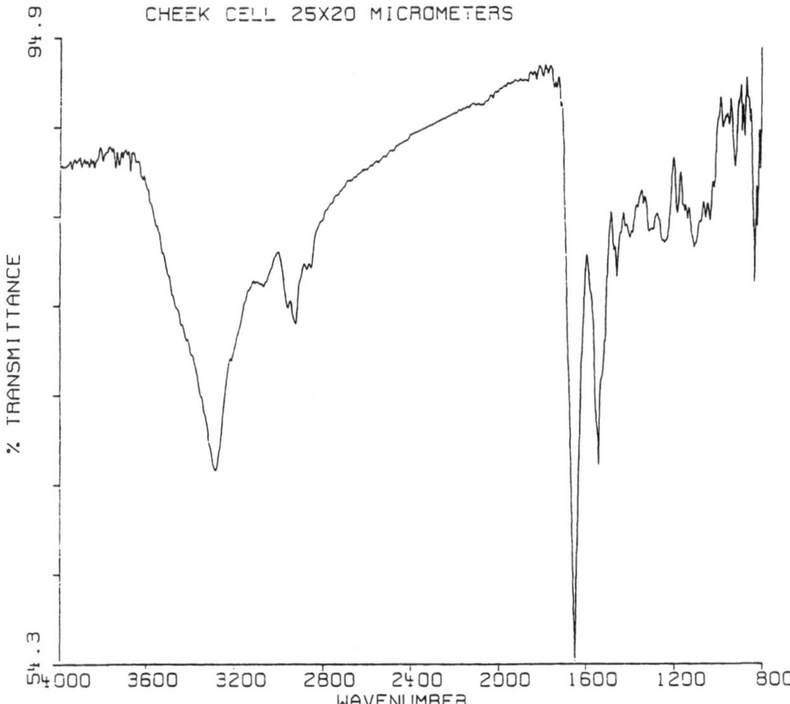

Figure 6.9 The infrared spectrum of an individual cheek cell obtained with an infrared microscope.

Gas chromatography has two limitations. First, it only works on molecules that can be volatilized at temperatures less than 300°C. Second, it provides chemical structure information in a very indirect manner. Standards of known composition must be run and the retention times compared to that of an unknown to determine chemical identity. If a standard's retention time is similar to that of an unknown peak, it may be assumed the two peaks contain the same molecule. However, this is not always a safe assumption since different molecules can have the same retention time. Also, if none of the standard's retention times match those of an unknown, no identification is possible.

Obtaining accurate chemical identification of the components in GC samples can be obtained by interfacing the GC to an FTIR. This gives rise to a new hyphenated technique called Gas Chromatography-Fourier Transform Infrared, which is abbreviated GC-FTIR or GC-IR. Figure 6.10 shows a sketch of a GC-FTIR instrument. The GC is on the left and contains an injector, column, oven, and a detector. There is also an oven temperature controller and a plotter to obtain a hard copy of the chromatogram. A heated transfer line carries the sample gas from the GC column to the GC-FTIR interface. It is

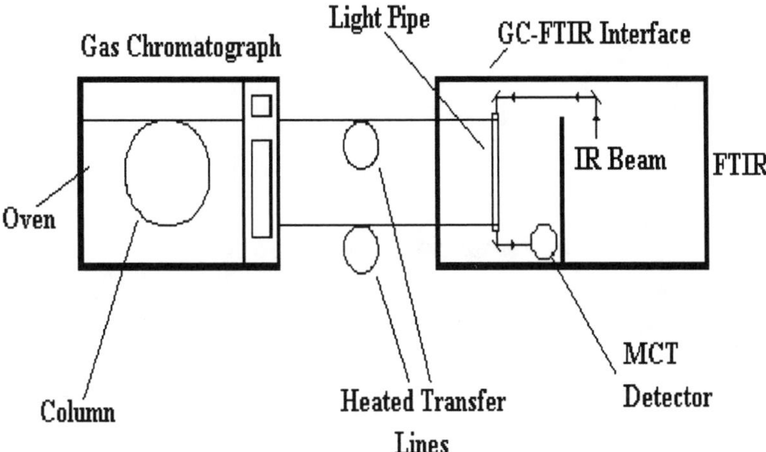

Figure 6.10 A diagram of a Gas Chromatography-Fourier Transform Infrared (GC-FTIR) instrument .

crucial that the transfer line be heated controllably. If the line is too cool, the sample will condense in the line, blocking it and preventing any sample from getting to the interface. After leaving the GC, the sample gases are swept into what is called a *lightpipe*, which is mounted inside the GC-FTIR interface. A lightpipe is a section of GC column the inside of which has been coated with gold. A diagram of a lightpipe is shown in Figure 6.11. A lightpipe is equipped with special fittings to avoid dead volume, i.e., space that is not constantly flushed with gas. The sample molecules are swept into one end of the lightpipe, flow down its length, then out of the lightpipe to return to the GC. The infrared beam is directed by a mirror through a hole in the side of the spectrometer to the GC-FTIR interface. A mirror in the interface focuses the IR beam on the lightpipe. There are KBr windows mounted on either end of the lightpipe to allow the infrared beam to pass. The infrared beam traverses the lightpipe by bouncing off the gold surfaces. Once the IR beam leaves the lightpipe, it is focused on a high sensitivity, liquid nitrogen cooled, narrow band MCT detector. A high sensitivity detector is essential since such a small amount of material is being detected.

The FTIR leaves the sample intact, so a GC chromatogram is also obtained in a GC-FTIR experiment. The chromatogram can be used to quantify the components in a mixture. The FTIR is much less sensitive than most GC detectors, so the FTIR will not detect as many sample peaks as the GC. Also, it takes some time (~5 seconds) for the sample to travel from the lightpipe to the GC detector. So, the retention times measured by the GC will be slightly longer than the retention times measured by the FTIR.

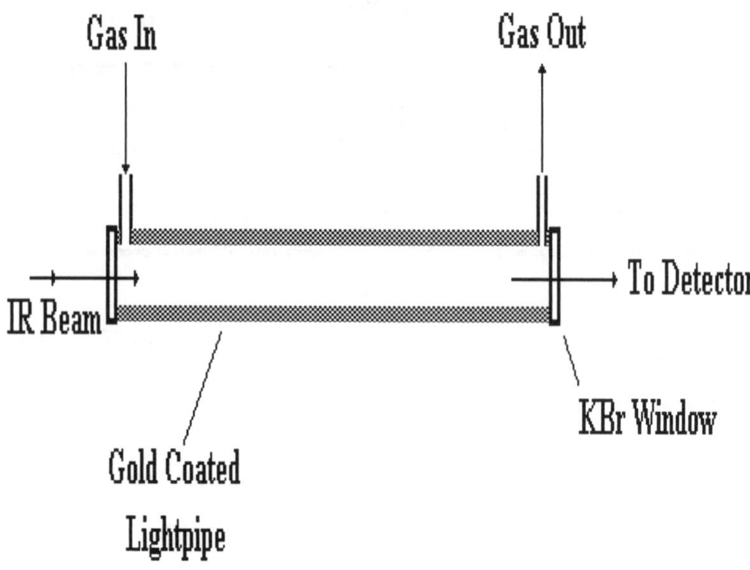

Figure 6.11 A diagram of a GC-FTIR lightpipe, the heart of a GC-FTIR interface.

Software Issues

The swiftness with which GC effluents are swept through the lightpipe poses a real challenge for FTIR computers and software. The FTIR has no way of knowing when there is sample gas in the lightpipe. Therefore, spectra must be taken continuously to obtain the spectrum of the sample gas while it is in the lightpipe. Often times several interferograms are added together, Fourier transformed, and saved to disk in what is known as a *scanset*. The generation of one scanset per second is common in GC-FTIR. During the course of a long GC run, thousands of scansets are obtained and saved. Only those scansets that were obtained when there was a sample in the lightpipe are of interest. The problem then becomes one of sorting through all the data to find the spectra of interest. Fortunately, there is a solution to this problem. In addition to collecting spectra, GC-FTIR software collects what are known as *infrared chromatograms*, which are simply plots of infrared absorbance versus time. Several examples of infrared chromatograms are shown in Figure 6.12. There are two basic types of infrared chromatograms. A *Gram-Schmidt chromatogram* is a plot of total infrared absorbance versus time, and tells the user exactly when an infrared absorbing species is in the lightpipe. The second type of infrared chromatogram is called a *functional group chromatogram*, which is a plot of the infrared absorbance in a specific wavenumber range versus time. These plots give information about the chemical composition of the species in the lightpipe as a function of time. The bottom chromatogram in Figure 6.12 is of the Gram-Schmidt variety. The top four chromatograms in the figure are functional group chromatograms. The wavenumber range for each of these chromatograms is noted to the right in the figure.

In a process called *peak editing,* the infrared chromatograms are examined, and spectra that were obtained when sample gas was in the lightpipe can be extracted. Most peaks in an infrared chromatogram are several seconds wide. If several scansets were obtained while a specific sample was in the lightpipe, these scansets can be added together to enhance the SNR of the spectrum (see Chapter 2 for a discussion on how adding scans together increases SNR). Once the spectra of relevant GC peaks are obtained, they can be searched against gas phase libraries to identify them. Since the GC purifies the sample, and gas phase spectra have many sharp bands, GC-FTIR library searches are often of high quality. This is what makes GC-FTIR so powerful: the ability to separate, identify, and quantitate a complex sample with one experiment.

An example of a real world GC-FTIR experiment is shown in Figures 6.12 through 6.14. As already stated, Figure 6.12 contains several infrared chromatograms of a GC-FTIR sample. Six components can be observed in the Gram-Schmidt chromatogram. All of the functional group chromatograms show six peaks each except for the functional group chromatogram from 1600 to 1800 cm^{-1}. This chromatogram shows only one peak at 6.19 minutes, which is the second component eluting from the column. The search results for the first eluting component (retention time of 5.48 minutes) are shown in Figure 6.13. The substance was identified as decane, a common straight chain hydrocarbon. The search results for the second eluting component, with the unique absorbance between 1600 and 1800 cm^{-1} (retention time 6.19 minutes), are shown in Figure 6.14. The substance was identified as a substituted cyclohexene (HQI = 0.08). The two best library hits are for stereoisomers of the same molecule. The HQI for the third library hit (0.28) is significantly different from the HQI for the two best hits. This means the compound is almost definitely the substituted cyclohexane. What set the 6.19 minute component apart was the presence of a double bond in its structure, which gives rise to a C=C stretch band between 1600 and 1700 cm^{-1}. This is why the 1600 to 1800 cm^{-1} functional group chromatogram contained a peak at 6.19 minutes.

The applications of GC-FTIR are wide and varied. Analysis of pesticides, herbicides, and polychlorinated biphenyls (PCBs) in environmental samples has been accomplished. The petroleum industry makes use of GC-FTIR to look at the constituents of gasoline, fuel oil, and other petroleum products. The chemical industry uses GC-FTIR to look at feed streams, intermediates, and final products to ascertain their composition and quality. The flavors and fragrances industry uses GC-FTIR to analyze the components of flavor oils, foods, and beverages that give rise to taste and smell. GC-FTIR can easily

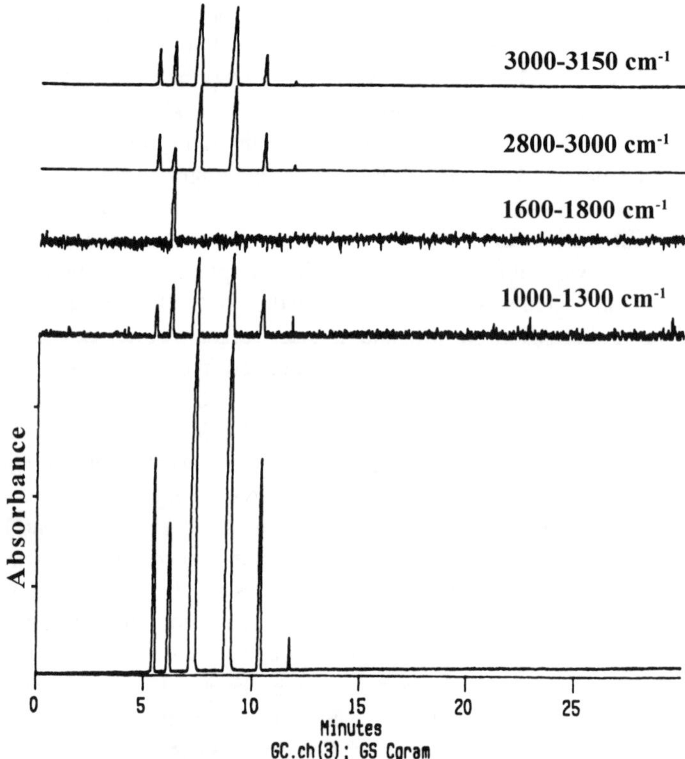

Figure 6.12 Examples of infrared chromatograms obtained during a GC-FTIR experiment. Bottom: A Gram-Schmidt chromatogram, which is a plot of total IR absorbance versus time. Top: Four different functional group chromatograms, which are plots of IR absorbance in a specific wavenumber range versus time. The wavenumber range for each chromatogram is noted on the right side of the figure.

distinguish between chemical isomers to provide the structure of the compound of interest. GC-MS can not distinguish between isomers, and this is one of its disadvantages compared to GC-FTIR. The rapidly expanding field of drug testing makes use of GC-FTIR to identify drugs of abuse in bodily fluids. All told, GC-FTIR has a wide range of applications, and the number expands with each passing year.

C. High Pressure Liquid Chromatography-Fourier Transform Infrared Spectroscopy (HPLC-FTIR)

High Pressure Liquid Chromatography (HPLC or LC) is an analytical tech-

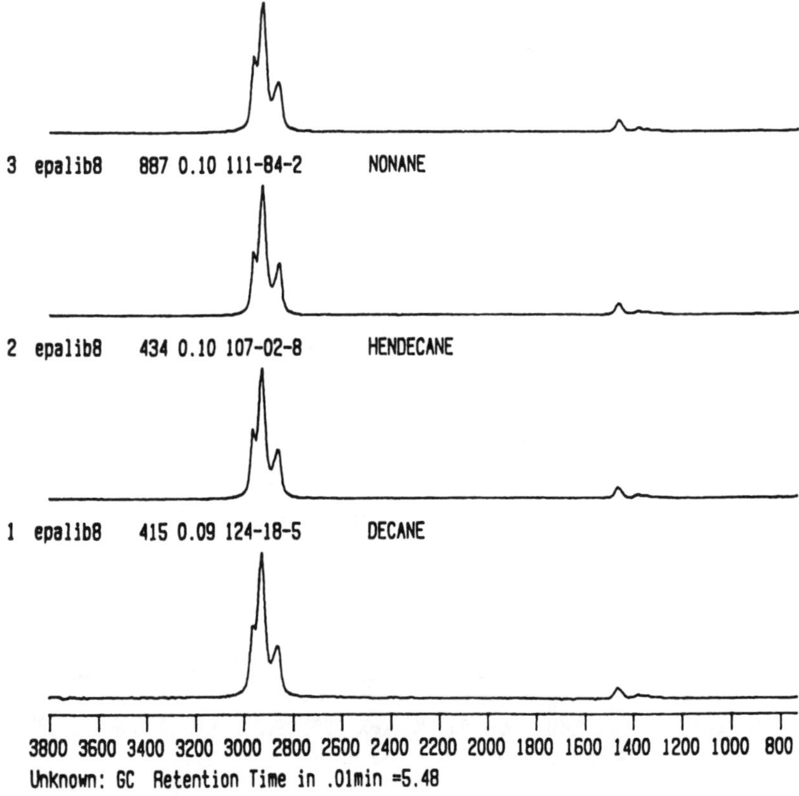

Figure 6.13 Library search results for the first component to come off the GC column, which had a retention timeof 5.48 minutes. The best library match is decane, which has a hit quality index of 0.09 (0.0 is a perfect hit in this system).

nique used to separate and quantitate the components in a complex mixture. A schematic diagram of an HPLC system is shown in Figure 6.15. The pump draws a solvent, also known as the mobile phase or *eluant*, out of a reservoir. Samples are dissolved in the eluant, and are introduced into the flowing eluant using an injector. The sample is pumped to a column which contains an adsorbent material called the stationary phase. The stationary phase often times consists of silica or coated silica. The molecules in the sample interact with the stationary phase, and elute at different points in time, depending on the strength of the interaction. Molecules that interact weakly with the stationary phase elute first; molecules that interact strongly elute last. The separation can be based on a molecule's polarity, size, and charge (to name a few of the many possible separation mechanisms). A detector is used to deter-

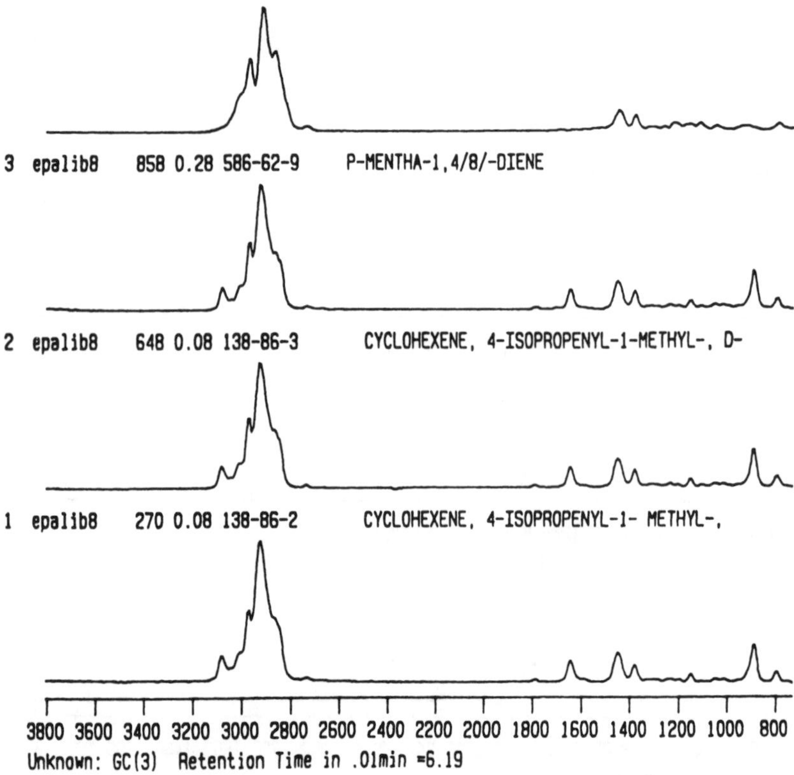

Figure 6.14 Library search results for the second component to come off the GC column, which had a retention time of 6.19 minutes. The best library match is a substituted cyclohexene, which has a hit quality index of 0.08 (0.0 is a perfect hit in this system).

mine when molecules come off the column. A plot of detector signal versus time is called a *chromatogram*, which is the fundamental measurement made in HPLC. Typical detectors use a sample's refractive index or ultraviolet-visible (UV/VIS) absorbance spectrum to measure an chromatogram. The response of these detectors is proportional to the amount of sample in the eluant, which is how quantitation is accomplished in HPLC. The problem with these detectors is that they provide little chemical information. They tell when molecules are eluting off the column, not what molecules are eluting.

By interfacing an HPLC to an FTIR, the components of complex mixtures can be separated and identified at the same time. LC-FTIR is different from GC-FTIR in that their separation mechanisms are different. Also, HPLC can be performed on any molecule that can be dissolved in the eluant, while GC is limited to molecules with boiling points below 300°C. Another major differ-

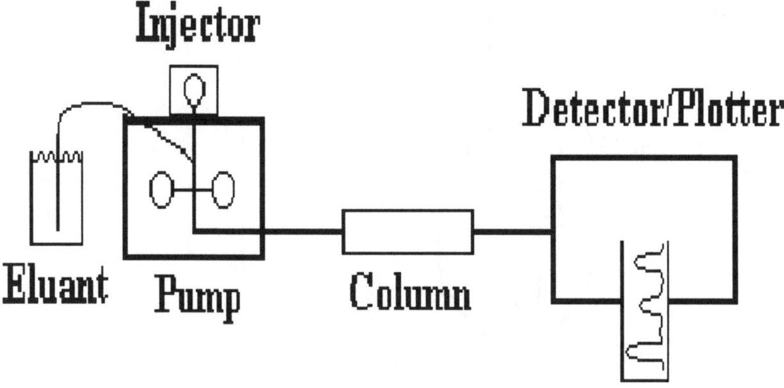

Figure 6.15 A diagram of a High Pressure Liquid Chromatography (HPLC) system.

ence between LC and GC, and the bane of LC-FTIR, is that solvents are used in HPLC. These solvents almost always absorb in the infrared, and interfere with the spectra of the molecules of interest.

Since the solvent causes problems in LC-FTIR, the best approach to performing this technique is to get rid of the solvent. There are several ways of doing this. The simplest solvent elimination method is to collect fractions from the LC run when the detector indicates there is sample coming off the column. The solvent can be evaporated or blown off, and the sample collected and its infrared spectrum obtained in a traditional way. Also, the HPLC eluant can drip into a DRIFTS cup loaded with KBr. The solvent evaporates, leaving behind the solute. The DRIFTS spectrum of the solute is obtained by placing the sample cup in a DRIFTS accessory.

The most exciting development of late in HPLC-FTIR is the introduction of devices that remove the LC solvent as soon as it emerges from the HPLC plumbing. A diagram of a typical device for this purpose is shown in Figure 6.16. The effluent from the HPLC flows through the inner tube in the drawing. Hot nitrogen gas flows through the outer tube, and is focused on the outlet of the inner tube. The warm nitrogen causes the solvent to evaporate as it exits the plumbing, leaving behind particles of solute. Organic or aqueous eluants can be evaporated in this way, and high flow rates of eluant can be handled by putting a splitter in-line, allowing only part of the fluid flow to pass to the solvent elimination device.

After evaporation, the solute particles fall upon a rotating germanium or metal mirror. Different HPLC peaks are deposited at different spots on the mirror. Subsequent to the HPLC run, the mirror is placed in a special accessory that mounts in the sample compartment of most FTIRs. A diagram of

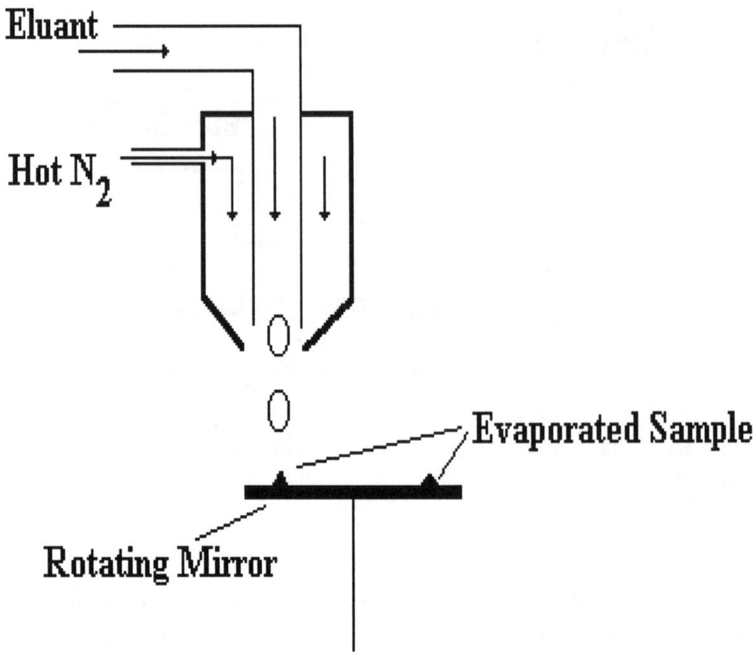

Figure 6.16 A diagram of a solvent removal system used in HPLC-FTIR.

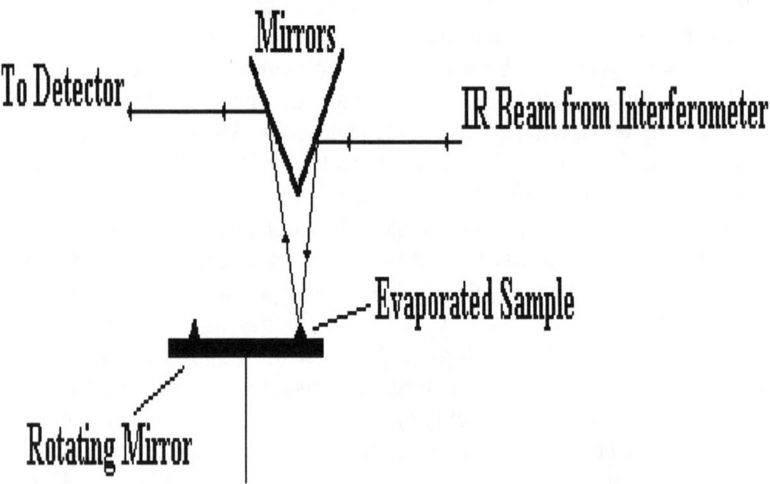

Figure 6.17 The FTIR accessory used to obtain spectra of evaporated HPLC samples.

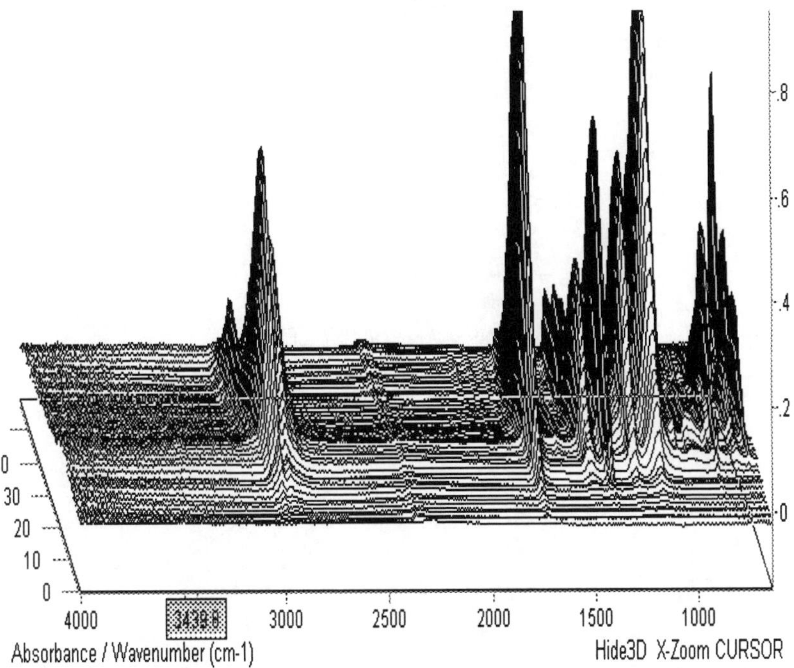

Figure 6.18 An example of data taken with an HPLC-FTIR. The X axis is in wavenumbers, the Y axis in absorbance, and the Z axis (pointing into the plane of the page) is in minutes. The plot represents FTIR spectra taken at different points in time during an HPLC run. The sample is a polystyrene/ polymethyl methacrylate copolymer.

this accessory is shown in Figure 6.17. The IR beam is focused on the surface of the mirror using a beam condenser, and the mirror is rotated so that different samples pass through the infrared beam at different points in time. The collection optics then send the IR beam to the FTIR's detector. If the mirror is rotated at the same speed during the HPLC and FTIR runs, a chromatographic peak and its corresponding spectrum will have the same time value. For example, if there is an HPLC peak at 7.5 minutes, the spectrum taken at 7.5 minutes will be of that peak. A commercially available HPLC-FTIR interface that makes use of this technology is built and sold by Lab Connections Inc. (Marlborough, MA). Software similar to the GC-FTIR software discussed above is used in LC-FTIR to acquire spectra quickly and to generate infrared chromatograms. Afterwards, peak editing is used to extract the spectra of interest, and library searches can be used to identify the components.

A problem with any HPLC-FTIR interface is that in LC separations, buffers and other materials are added to the eluant. If these materials do not have a

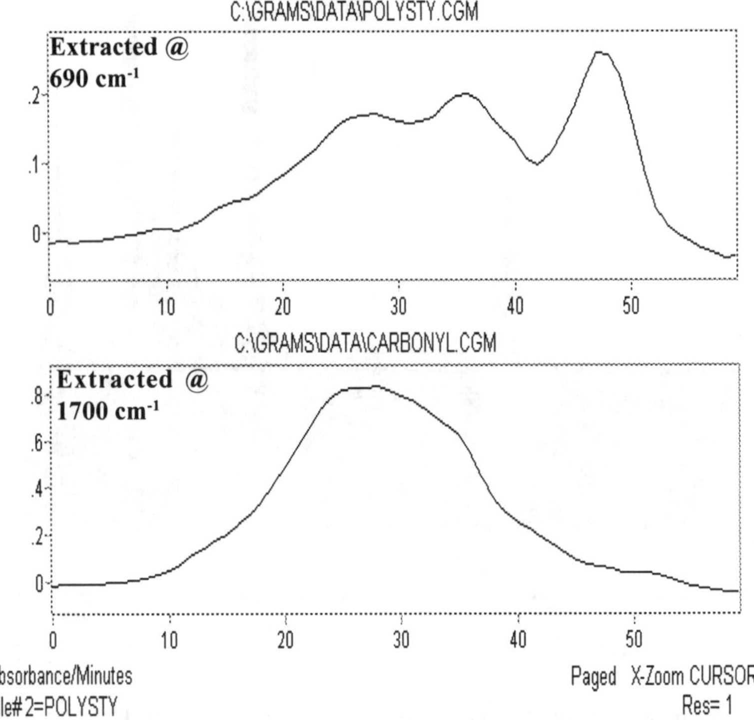

Figure 6.19 Examples of functional group HPLC-FTIR chromatograms extracted from the same data set. Bottom: A plot of infrared absorbance at 1700 cm^{-1} versus time. Top: A plot of infrared absorbance at 690 cm^{-1} versus time.

low enough boiling point, they will not evaporate after coming out of the plumbing. As a result, the buffers will contaminate the sample and contribute unwanted bands to the infrared spectrum of the sample. Sometimes the spectrum of the buffer can be subtracted from sample, but as discussed in Chapter 3, subtractions do not always work properly. Volatile buffers such as ammonium acetate will evaporate and not contaminate the sample. However, redeveloping a separation method to use a new buffer can be time consuming.

An example of data obtained with an HPLC-FTIR is shown in Figure 6.18. The sample was a polystyrene/polymethyl methacrylate copolymer. The copolymer was dissolved in a solvent, then passed through a gel permeation chromatography (GPC) column, where the separation mechanism is based on molecular size and weight. In GPC the largest molecules elute from the column first, and the smallest molecules elute last. So, the retention time of a chromatographic peak is related to molecular weight. In Figure 6.18 the X axis is in cm^{-1}, the Y axis is in absorbance, and the Z axis (which points into the plane of the page) is in minutes. The plot shows infrared spectra taken at different points in time during a GPC run.

Recall that the examination of infrared chromatograms can tell a lot about a sample. An example of two functional group chromatograms extracted from the data seen in Figure 6.18 is shown in Figure 6.19. The bottom chromatogram in the figure is a plot of the absorbance at 1700 cm^{-1} vs. time. This absorbance is due to the carbonyl stretch of the methyl methacrylate (MMA) groups. So, this plot represents how the amount of MMA in the copolymer changed with molecular weight. The chromatogram at the top of Figure 6.19 is of the absorbance at 690 cm^{-1} vs. time. This absorbance is due to the benzene ring of the polystyrene, and this plot represents how the amount of styrene in the copolymer changed with molecular weight. The chromatograms have different structures, indicating that the distribution of styrene in the polymer has a different dependence on molecular weight than the distribution of the methyl methacrylate. In essence, the lower molecular weight polymer molecules (that come off late in the HPLC run) are enhanced in polystyrene compared to the higher molecular weight molecules. This is just one example of the interesting things that can be discovered using HPLC-FTIR.

References

1. P. Wong, R. Wong, and M.F.K. Fung, *Appl. Spec.* **47**(1993)1058

Bibliography

Robert Messerschmidt and Matthew Harthcock, Eds., *Infrared Microspectroscopy: Theory and Applications*, Marcel Dekker, New York, 1988.

Howard Humecki, Ed., *Practical Guide to Infrared Microspectroscopy,* Marcel Dekker, New York, 1995.

K. Krishnan and S. Hill, FTIR Microsampling Techniques, Bio-Rad/Digilab FTIR Applications Note No. 73, Bio-Rad Corporation, Cambridge MA, 1990.

J. Ferraro and K. Krishnan, Eds., *Practical Fourier Transform Infrared Spectroscopy*, Academic Press, New York, 1990.

J. Durig, Ed., *Applications of FTIR Spectroscopy*, Elsevier, Amsterdam, 1990.

Glossary

This glossary contains definitions of many important FTIR terms. Many of the terms listed here appeared *in italics* in the body of the book. Words that appear in italics in the glossary are defined elsewhere in this section.

100% Line - is calculated by ratioing two background spectra taken under identical conditions. Ideally, the result is a flat line at 100% transmittance. The slope and noise of 100% lines are measured to ascertain the quality of spectra and the health of an instrument.

Absorbance - units used to measure the amount of infrared radiation absorbed by a sample. Absorbance is commonly used as the Y axis unit in *infrared spectra*. Absorbance is defined by *Beer's law*, and is linearly proportional to concentration. This is why spectra plotted in absorbance units should be used in quantitative analysis.

Air Bearing - a device used in some interferometers to reduce friction between the moving mirror and its housing. In this particular device, the moving mirror and its shaft are supported on a cushion of air.

Air Cooled Source - a type of infrared source used in infrared spectrometers. This source typically consists of coil of wire or a small ceramic piece through which electricity passes, causing it to give off heat. It is said to be "air cooled" since air currents in the instrument affect its temperature.

Analog-to Digital Converter (ADC) - is used to convert an interferogram signal from volts (the analog signal) into a series of base 2 numbers (a digital signal), the language that a computer understands. ADCs are characterized by the number of "bits" that the digitized output contains.

Analyte - the molecule of interest when performing a quantitative analysis.

Angular Divergence - the spreading out of an infrared beam as it travels through the FTIR. Angular divergence contributes to noise in high resolution spectra.

Apodization Functions - are multiplied times an interferogram to reduce the amount of *sidelobes* in a spectrum. Different types of apodization functions include boxcar, triangle, Beer-Norton, Happ-Genzel, and Bessel. The use of apodization functions unavoidably reduces the resolution of a spectrum.

ATR - stands for Attenuated Total Reflectance, and is a *reflectance sampling* technique. In ATR, infrared radiation impinges on a prism of infrared transparent material of high refractive index. Because of *internal reflectance,* the light reflects off the surface of the crystal at least once before leaving it. The infrared radiation sets up an *evanescent wave* which penetrates a small distance above and below the crystal surface. Samples brought into contact with the surface will absorb the evanescent wave giving rise to an infrared spectrum. This sampling technique is useful for liquids, polymer films, and semisolids.

Background Spectrum - a single beam spectrum acquired with no sample in the infrared beam. The purpose of a background spectrum is to measure the contribution of the instrument and environment to the spectrum. These effects are removed from a sample spectrum by rationg the *sample single beam* spectrum to the background spectrum.

Baseline Correction - a spectral manipulation technique used to correct spectra with sloped or curving baselines. The user must draw a function parallel to the baseline, then this function is subtracted from the spectrum.

Beer's Law - the equation that relates the *absorbance* of a sample to its concentration. Its form is $A = \varepsilon lc$ where the A is stands for absorbance, ε for absorptivity, l is for pathlength, and c is for concentration. Beer's law is the equation used in FTIR quantitative analysis to perform *calibrations* and to predict unknown concentrations.

Calibration - the process in quantitative analysis by which the peak heights and areas in a spectrum are correlated with the concentrations of *analytes* in *standards.* After calibration, the concentration of the analyte in unknown samples can be predicted.

Calibration Curve - a plot of absorbance versus concentration used in a *calibration.* If the plot is linear, it means *Beer's law* has been followed, and that the plot can be used to predict the concentration of unknown samples.

Capillary Thin Film - a *transmission sampling* technique used to obtain spectra of liquids. Typically, a drop of liquid is placed between two KBr windows, and the windows and sample are placed directly into the infrared beam. The capillary action of the liquid holds the two windows together, hence the name of the technique.

Cast Films - a *transmission sampling* technique used to analyze polymer films. The polymer is dissolved in a solvent, and the solution is evaporated onto a KBr window giving a polymer film. The window/film combination is then placed directly in the infrared beam.

Centerburst - the sharp, intense part of an *interferogram.* The size of the centerburst is directly proportional to the amount of infrared radiation striking the detector.

Cepstrum - the result obtained when a reverse *Fourier transform* is performed on a spectrum. Cepstrums are used in the *deconvolution* process.

Chromatogram - a plot of detector response versus time. The response is typically proportional to amount of material coming out of a gas or liquid *chromatograph*. These devices are interfaced to FTIRs to give *hyphenated techniques.*

Chromatograph - a device used to separate complex mixtures into their components. Chromatographs measure *chromatograms*, which are plots of detector response versus time. Chromatographs can be used to quantify the concentration of molecules in a sample.

Coadding - the process of adding *interferograms* together to achieve an improvement in *signal-to-noise* ratio.

Condenser - the optic in a microscope that focuses light onto the sample.

Constructive Interference - a phenomenon that occurs when two waves occupy the same space and are in phase with each other. Since the amplitudes of waves are additive, the two waves will add together to give a resultant wave which is more intense than either of the individual waves. This phenomenon causes the *centerburst* seen in *interferograms*.

Cross Validation - when using *factor analysis*, the process by which the optimal number of factors is determined. Essentially, *calibrations* are performed by leaving specific *standards* out of the calculation, then actual and calculated concentrations are compared.

Curvefitting - a method used to try and determine the number, shape, position, width, and height of a group of infrared bands that overlap to give one broad band. The technique involves making initial estimate of the parameters, then using a *least squares* fitting routine to optimize the parameters.

Deconvolution - a way of mathematically enhancing the resolution of a spectrum to visualize spectral features that overlap to give a broad band. The calculation involves the use of *cepstrums.*

Depth of Penetration - in *ATR*, a measure of how far infrared radiation penetrates into the sample. More precisely, it is the depth at which the *evanescent wave* has decreased to 37% of its original value. Depth of penetration is affected by a number of things including the wavenumber of infrared radiation, the refractive index of the ATR crystal, and the angle of incidence of the light impinging on the sample.

Depth Profiling - the ability of some sampling techniques (e.g., *ATR* and *photoacoustic spectroscopy*) to obtain spectra from different depths within a sample nondestructively. This allows the change in composition with depth to be studied.

Derivative - in mathematics, a plot of the slope of a function versus its X axis values. In FTIR, the derivative of a spectrum gives sharp features at the X axis value of greatest absorbance. Thus, derivatives are used in peak picking and *library searching*.

Destructive Interference - a phenomenon that occurs when two waves occupy the same space. Since the amplitudes of waves are additive, if the two waves are out of phase with each other, the resultant wave will be less intense than either of the individual waves.

Diamond Anvil Cell - a device used to prepare samples for *transmission sampling* by *infrared microscopy*. The cell consists of two diamonds with flat faces. The sample is placed on one diamond face, and the second diamond face is brought into contact with the sample to squish it. The entire assembly is then placed in the infrared beam of the *infrared microscope*.

Diffuse Reflectance - the phenomenon that takes place when infrared radiation reflects off a rough surface. The light is transmitted, absorbed, scattered, and reflected by the surface. The light approaches the surface from one direction, but the diffusely reflected light leaves the surface in all directions. A *reflectance sampling* technique known as *DRIFTS* is based on this phenomenon.

Dispersive Instruments - *infrared spectrometers* that use a grating or prism to disperse infrared radiation into its component wavenumbers before detecting the radiation. This type of instrument was dominant before the development of FTIR.

DRIFTS - stands for Diffuse Reflectance Fourier Transform Infrared Spectroscopy, a *reflection sampling* technique that makes use of the phenomenon of *diffuse reflectance*, and is used primarily on powders and other solid samples.

Electric Vector - the electric part of electromagnetic radiation (light). It is the electric vector that interacts with molecules to give rise to infrared spectra.

Evanescent Wave - in *ATR,* the standing wave of radiation set up in the ATR crystal. The evanescent wave penetrates beyond the crystal surface into any sample brought into contact with the surface. As a result, the infrared spectrum of the sample can be obtained.

External Standards - a technique of quantitative analysis where the *standards* are run at a different point in time than the unknown samples.

Factor Analysis - a method of multicomponent quantitative analysis.

Fourier Transform - the calculation performed on an interferogram to turn it into an *infrared spectrum*. The calculation involves a mathematical integral.

Fourier Transform Infrared (FTIR) - a method of obtaining *infrared spectra* by first measuring the *interferogram* of the sample using an *interferometer*, then performing a *Fourier transform* on the interferogram to obtain the spectrum.

Full Spectrum Search - in *library searching*, the use of entire spectra when comparing unknown and library spectra. The advantage of this method is that the use of all the spectral data points gives a more accurate comparison.

Functional Groups - structural fragments within a molecule that have unique reactivity or properties. For instance, the carbonyl group in acetone is an example of a functional group. The main use of *infrared spectroscopy* is the detection of specific functional groups within molecules.

Full Width at Half Height - a way of measuring the width of an infrared band. The width of the band is measured at half the maximum absorbance value of the band. The peak width at full height of some gas phase molecules is used to measure the *resolution* that an *infrared spectrometer* has attained.

Functional Group Chromatogram - in *GC-FTIR*, a plot of the absorbance in a specific wavenumber range versus time. Typically, the wavenumber range is chosen to correspond the absorbance of a specific *functional group.*

GC-FTIR - stands for Gas Chromatography-Fourier Transform Infrared, a type of *hyphenated technique*. GC-FTIR involves interfacing a gas *chromatograph* (GC) to an FTIR to obtain the spectra of sample molecules as they come out of the GC. By combining the two techniques, complex mixtures can be analyzed quickly.

Gram-Schmidt Chromatogram - in *GC-FTIR,* a plot of total infrared absorbance versus time. This chromatogram is used to determine how many sample components were detected by the FTIR.

Heat and Pressure Films - a *transmission sampling* technique used to obtain spectra of polymers. Polymer samples are heated under pressure until they flow and form a thin film. The film is then placed directly in the infrared beam.

Hit Quality Index (HQI) - in *library searching*, the number that shows how closely matched a library spectrum is to an unknown spectrum.

HPLC-FTIR - stands for High Pressure Liquid Chromatography-Fourier Transform Infrared. This is a *hyphenated technique* where a high pressure liquid *chromatograph* is interfaced to an FTIR, which detects molecules as the leave the chromatograph.

Hyphenated Techniques - when an FTIR is interfaced to another instrument that also performs chemical or physical analyses, a hyphenated technique is born. The name derives from the fact that the new technique is usually abbreviated with a hyphen and the letter FTIR, such as *GC-FTIR*. By interfacing FTIRs to other instruments, more information about a sample can be obtained more quickly than using the two instruments to analyze the sample separately.

Infrared Chromatogram - a plot of infrared absorbance versus time. The *hyphenated techniques GC-FTIR* and *HPLC-FTIR* generate infrared chromatograms. Infrared chromatograms are used to determine at what points in time the FTIR detected something coming out of the *chromatograph. Functional group chromatograms* and *Gram-Schmidt chromatograms* are examples of infrared chromatograms.

Infrared Microscope - a microscope specially designed to handle visible and infrared radiation. The microscope is interfaced to an FTIR and used to visually examine and take the infrared spectrum of small (less than 250 microns in diameter) samples.

Infrared Micrcoscopy - also known as infrared microspectroscopy. The technique of using an *infrared microscope* to obtain the *infrared spectrum* of microscopic samples.

Infrared Radiation - the portion of the electromagnetic (light) spectrum from 10 to 14,000 cm^{-1}. This type of light is higher in wavenumber than radio and microwaves, but is smaller in wavenumber than visible light. Infrared radiation is the same thing as heat.

Infrared Spectrometer - an instrument that is used to obtain the *infrared spectrum* of a sample. The instrument typically consists of a source of infrared radiation, a sample compartment to allow the radiation to interact with a sample, a means of determining the intensity of radiation as a function of wavenumber, and a way of displaying the spectrum of the sample.

Infrared Spectroscopy - the study of the interaction of infrared radiation with matter.

Infrared Spectrum - a plot of measured infrared intensity versus wavenumber. The features in an infrared spectrum correlate with the presence of *functional groups* in a molecule, which is why infrared spectra can be interpreted to determine molecular structure.

Instrument Response Function - the portion of a *background spectrum* due to the instrument. The instrument's components, such as mirrors, detector, and beamsplitter all contribute features to the instrument response function.

Interferogram - a plot of infrared detector response versus *optical path difference.* The fundamental measurement obtained by an FTIR is an interferogram. Interferograms are *Fourier transformed* to give *infrared spectra.*

Interferometer - an optical device that causes two beams of light to travel different distances (*optical path difference*). The light beams are combined allowing *constructive interference* and *destructive interference* to take place. By changing the optical path difference using the interferometer, an *interferogram* is measured.

Internal Reflection - the phenomenon whereby light passing through material of high refractive index reflects off the surface of the material rather than passing out of it. Internal reflection takes place in fiber optics and in *ATR.*

Internal Standards - a technique used in quantitative analysis. A known amount of a known substance called the "internal standard" is added to all *standards* and unknowns. The *calibration curve* is plotted with the ratio of the *analyte's* absorbance to the internal standard absorbance on the Y axis. By ratioing the two absorbances measured at the same time, fluctuations in the instrument or sampling that may have affected absorbance values are cancelled out.

K Matrix - a multicomponent quantitative analysis technique, also known as "classical least squares" or CLS. In this method, absorbance is expressed as a function of concentration, and a *least squares fit* is performed to obtain the *calibration.*

KBr Pellet - a *transmission sampling* technique most commonly used on powders and other solids. The technique involves grinding the sample and KBr, diluting the sample in the KBr, then pressing the mixture to produce a transparent pellet. The pellet is then placed directly in the infrared beam.

Kramers-Kronig Transform - a mathematical calculation performed on *specular reflectance spectra* to get rid of *restsrahlen.* The result of the calculation is an "n spectrum", which is a plot of refractive index versus wavenumber, and a "k spectrum" which is the true absorbance spectrum of the sample.

KRS-5 - the tradename for thallium bromide iodide. It is commonly used as an *ATR* crystal. It is a very toxic substance, and should be handled with great care.

Kubelka-Munk Units - units used to measure the intensity of diffusely reflected light. The Kubelka-Munk equation relates Kubelka-Munk units to the concentration and scattering factor of a sample. The scattering factor is determined by the particle size, shape,and packing density of the sample, and can be difficult to control. Kubelka-Munk units should be used when performing quantitative *DRIFTS* analyses.

Least Squares Fit - a technique used to develop mathematical models of reality. By definition, a least squares fit will provide the best model available. In FTIR, least squares fits are used in multicomponent quantitative analysis to model the correlation between infrared spectra and concentration (*calibration*).

Library Searching - a process in which an unknown spectrum is compared to a collection of known spectra kept in a spectral library. The comparison gives a number called the *hit quality index* which represents how closely related two spectra are to each other. If a match is of high quality, it is possible to identify an unknown sample using library searching.

Mechanical Bearing - a device used in some *interferometers* to reduce friction between the moving mirror and its housing. In this particular device, the moving mirror and its shaft are typically held in place by ball bearings.

Microscope Mapping - using an *infrared microscope* to obtain spectra at different points in a sample. Often times a special computer controlled microscope stage is used to obtain spectra at precise spatial intervals. The result of microscope mapping is a *molecular map*, which is a plot of absorbance at a specific wavenumber versus position on the sample.

Mid-Infrared - infrared radiation between 400 and 4000 cm^{-1}.

Mirror Displacement - the distance that the mirror in an *interferometer* has moved from *zero path difference.*

Molecular Maps - plots of absorbance at a specific wavenumber versus location on a sample. Usually obtained using *microscope mapping*.

Monochromator - a device used in *dispersive instruments* to separate an infrared light beam into its component wavenumbers. A grating or prism is used for this purpose.

Mull - a *transmission sampling* technique where the sample is ground, then dispersed in an oil or *mulling agent*. The oil/sample mixture is then sandwiched between two KBr windows and placed in the infrared beam.

Mulling Agent/Mulling Oil - oil that is added to a ground sample for the preparation of *mulls*.

Multiplex (Felgett) Advantage - is an advantage of FTIR compared to dispersive instruments. It is based on the fact that in an FTIR all the wavenumbers of light are detected at once.

Normalized - the process of dividing all the absorbance values in a spectrum by the largest absorbance value. This resets the Y axis scale from 0 to 1. Normalization is often performed on spectra before library searching.

Objective - the optic in a microscope that collects light after it has interacted with the sample.

Optical Path Difference - the difference in distance that two light beams travel in an *interferometer*.

P Matrix - a multicomponent quantitative analysis technique, also known as "inverse least squares" or ILS. In this method, concentration is expressed as a function of absorbance, and a *least squares fit* is performed to obtain the *calibration.*

Peak Editing - a process typically performed after a *GC-FTIR* or *HPLC-FTIR* run. *Infrared chromatograms* are viewed to determine when sample molecules were detected by the FTIR. Then, the spectra acquired during the time when a sample was detected are *coadded* to improve *signal-to-noise ratio*.

Peak-to-Peak Noise - a noise measurement often made on a *100% line*. It is measured as the difference between the lowest and highest noise value in a specific wavenumber range. When obtained under controlled conditions, peak-to-peak noise is an excellent measure of spectrum quality and instrument health.

Photoacoustic Spectroscopy (PAS) - detecting a sample's infrared spectrum by "listening" to the sound made when the sample absorbs infrared radiation. A highly sensitive microphone is used as a detector, and the spectra are similar to absorbance spectra. This technique can be used for quantitative analysis.

Reference Spectrum - in *spectral subtraction*, the spectrum of a substance that is subtracted from the spectrum of a mixture (*sample spectrum*). Often, the reference spectrum is of a solvent.

Reflectance Sampling - a method of obtaining infrared spectra by bouncing the infrared beam off of the sample.

Reflection-Absorption - a *reflection sampling* technique used on thin films coated on shiny metal surfaces. The infrared beam passes through the film, reflects off the metal, then passes through the film a second time before reaching the detector. This technique is also known as "double-transmission".

Resolution - a measure of how well an *infrared spectrometer* can distinguish spectral features that are close together. For instance, if two features are 4 cm^{-1} apart and can be discerned easily, the spectrum is said to be at least 4 cm^{-1} resolution. Resolution in FTIR is determined by *optical path difference*.

Reststrahlen - *derivative* shaped spectral features that occasionally appear in *specular reflectance* spectra. Restsrahlen is caused by how a sample's refractive index changes in the vicinity of an absorbance band. The *Kramers-Kronig transform* can be used to remove restsrahlen from spectra.

Sample Single Beam Spectrum - a *single beam spectrum* obtained with a sample in the infrared beam. These spectra are typically ratioed against *background spectra* to obtain *absorbance* or *transmittance* spectra.

Sample Spectrum - in *spectral subtraction*, the spectrum of a mixture from which the *reference spectrum* is subtracted.

Sampling Depth - in *photoacoustic spectroscopy (PAS)*, the depth in the sample from which 63% of the photoacoustic signal is measured.

Scan - the process of measuring an interferogram with an FTIR. Typically, this involves moving the mirror in the *interferometer* back and forth once.

Sealed Liquid Cells - a *transmission sampling* technique used to obtain the spectra of liquids. The cell consists of two KBr windows held apart a fixed distance by a gasket. The cell is filled with liquid then placed in the infrared beam.

Search Algorithm - in *library searching*, the mathematical calculation used to compare two spectra and produce a *hit quality index*, which measures how closely matched two spectra are to each other.

Search Report - the end product of a *library search.* A search report ranks the quality of library matches using the *hit quality index* then presents these results in a table. Most software packages allow visual comparison of the unknown spectrum and the best library matches.

Sidelobes - spectral features that appear to the sides of an absorbance band as undulations in the baseline. Sidelobes are caused by having to truncate an *interferogram*,and can be removed from a spectrum by multiplying the spectrum's interferogram times an *apodization function.*

Signal-to-Noise Ratio (SNR) - the ratio of signal in a spectrum, usually measured as the intensity of an absorbance band, to noise measured at a nearby point in the baseline. SNR is a measure of the quality of a spectrum, and can be used to ascertain the quality of an *infrared spectrometer* if it is measured under controlled conditions.

Single Beam Spectrum - the spectrum that is obtained after *Fourier transforming* an *interferogram.* Single beam spectra contain features due to the instrument, the environment, and the sample (if there is one in the beam)

Smoothing - a spectral manipulation technique used to reduce the amount of noise in a spectrum. It works by calculating the average absorbance (or transmittance) of a group of data points called the "smoothing window", and plotting the average absorbance (or transmittance) versus wavenumber. The size of the smoothing window determines the number of data points to use in the average, and hence the amount of smoothing. The "smoothing algorithm" determines how the average is calculated.

Spectral Subtraction - a spectral manipulation technique where the absorbances of a *reference spectrum* are subtracted form the absorbances of a *sample spectrum.* The idea is to remove the bands due to the *reference* material from the sample spectrum. This is done by simply calculating the difference in absorbance between the two spectra, then plotting this difference versus wavenumber. The reference spectrum is often multiplied times a *subtraction factor* so that the reference material bands subtract out properly.

Specular Reflectance - the type of reflectance that takes place off smooth, shiny surfaces, such as that of mirrors. By definition, in specular reflectance the angle of incidence of light equals the angle of reflectance of light. Specular reflectance, as a *reflection sampling* technique, can be used to obtain infrared spectra.

Split Mulls - the technique of using two different *mulling oils*, namely Nujol and Fluorolube to obtain two *mulls* of the same sample. These two oils are transparent in different wavenumber ranges. By splicing the spectra of the two mulls together, a spectrum free of most *mulling oil* absorbances can be obtained.

Standards - in quantitative analysis, samples that contain known concentrations of the *analyte*. The absorbance of these samples is measured and then used in a *calibration*.

Subtraction Factor - in *spectral subtraction*, a number that is multiplied times the *reference spectrum* before it is subtracted from the *sample spectrum*. The purpose of the subtraction factor is to match the absorbances of the reference material in the two spectra so the reference bands subtract out cleanly. Also known as the "scale factor".

Subtrahend - see *reference spectrum*.

Thermal Wave - in *photoacoustic spectroscopy*, the heat deposited in a sample due to infrared absorption that travels towards the surface of the sample. It is caused by the fact that heat is conducted from areas of high temperature to areas of low temperature.

Throughput Advantage (Jacquinot Advantage) - an advantage of FTIR over *dispersive instruments* due to the fact that in FTIR all the radiation strikes the detector at once, enhancing *signal-to-noise ratio*.

Transmission Sampling - a sampling method where the infrared beam passes through the sample before it is detected. Samples are typically diluted or flattened to adjust the absorbance values to a measurable range.

Transmittance - a unit used to measure the amount of infrared radiation transmitted by a sample. It is often used as the Y axis unit in *infrared spectra*. Transmittance is <u>not</u> linearly proportional to concentration, and spectra plotted in these units should not be used for quantitative analysis.

Water Cooled Source - an infrared source whose temperature is controlled by a flow of water or other cooling liquid.

Wavelength - the distance between adjacent crests or troughs of a light wave.

Wavenumber - is defined as 1/*wavelength*. The units of wavenumbers are cm^{-1}, and are most commonly used as the X axis unit in *infrared spectra*.

Wings - the portion of an *interferogram* where there is little or no intensity.

Also, an excellent television sitcom currently airing on Tuesday nights.

Zero Path Difference - the *mirror displacement* at which the *optical path difference* for the two beams in an *interferometer* is zero.

Index

D

E

F

G

J

K

L

M